Integer Sequences

Masum Billal · Samin Riasat

Integer Sequences

Divisibility, Lucas and Lehmer Sequences

 Springer

Masum Billal
Dhaka, Bangladesh

Samin Riasat
Toronto, ON, Canada

ISBN 978-981-16-0572-7   ISBN 978-981-16-0570-3   (eBook)
https://doi.org/10.1007/978-981-16-0570-3

This Springer imprint is published by the registered company Springer Nature Singapore Pte Ltd.
The registered company address is: 152 Beach Road, #21-01/04 Gateway East, Singapore 189721,
Singapore

Dedicated to Morgan Ward

Preface

It gives us great pleasure that Springer Nature is publishing this book. Our initial purpose was to popularize the theories in this text within the mathematical olympiad community seeing how useful they are for solving many olympiad problems. However, we soon realized that this alone cannot be the sole purpose of the book as it covers areas that can also be classified as research material. We hope that the book will be able to help people on both fronts.

The Fibonacci sequence is one of the most popular topics of study in mathematics. Lucas sequences are a generalization of the Fibonacci sequence. In this book, we discuss Lucas sequences and their generalizations: Lehmer sequences, divisibility sequences, generalized binomial coefficients, and integer sequences with exponent lifting property. Our goal is to discuss theories that can be derived using mostly elementary means. Some notions have been borrowed from linear and abstract algebra, but they have been kept in check. We will not be using any particular idea that screams abstract algebra. Rather, we will use some basic notions to help develop our theories. Most results in this book are not usually found in common number theory texts. The theories discussed in this book are very interesting yet not as popular as one would hope. Considering the impact and usefulness of these theories, it seems unfortunate that theories like divisibility sequences are not part of regular studies and olympiad mathematics.

We had a very hard time finding proper references for some topics. For example, double modulus was found only in Venkov [2]. Fortunately, an English translation of the book was available, see Venkov [3]. Considering that the theory of double modulus has been known since the time of Dedekind, it is a shame that almost no common elementary text today covers this topic. Another gem we want to mention is the *Schönemann theorems*. We have yet to find proper references for Schönemann's theorems, especially his second theorem. The only references we could find of these theorems are in Ward's papers, e.g., Ward [4]. Apparently, Ward was a big fan of these theorems and often applied them in his work. Moreover, judging by papers of Engstrom [1] and others we are convinced they too knew these results well. For example, Engstrom [1] uses a factorization similar to that of Schönemann's in his paper. But today there is no record of these results either online or in any common books that we know of.

This book is dedicated to Morgan Ward due to his enormous contribution to the subject. Ward wrote 33 papers on recurring series, which is closely related to both divisibility sequences and Lucas sequences. We do not know for sure what the motivation was for Ward to study these sequences in such detail. For the first author, the initial motivation to study these topics was to find out when an integer sequence has integer binomial coefficients. For the second author, they were the many interesting divisibility properties of Fibonacci-like sequences. Ward's motivation might have been similar. A lot of the results and papers referenced are available only because of Ward. Although many original proofs have been replaced by modern notions and notations, it seems quite fitting that this book be dedicated to him.

Finally, we would like to express our gratitude to the kind referees whose opinions helped improve the original manuscript significantly. Nevertheless, in the case of any typo or error in results, the authors bear full responsibility and request to be contacted via email at billalmasum93@gmail.com or sriasat@uwaterloo.ca.

Dhaka, Bangladesh   Masum Billal
Toronto, Canada   Samin Riasat
April 2021

References

1. H.T. Engstrom. Periodicity in Sequences Defined by Linear Recurrence Relations. in *Proceedings of the National Academy of Sciences 16.10* (1930), pp. 663–665. https://doi.org/10.1073/pnas.16.10.663.
2. B.A. Venkov. Elementarnaia Teoriia Chisel. The main edition of general technical and techno-theoretical literary, 1937.
3. B.A. Venkov. Elementary Number Theory. Wolters-Noordhoff, 1970.
4. Morgan Ward. The Arithmetical Theory of Linear Recurring Series. Trans. Amer. Math. Soc. 35.3 (1933), pp. 600–628. ISSN: 00029947. http://www.jstor.org/stable/1989851.

Who This Book Is For

This book is written for pretty much anybody interested in Fibonacci-like sequences. We have tried to write the book in a way that not only undergraduates but people with a focus on mathematical olympiads can also read the book with little effort. As for pre-requisites, we have tried our best to keep the book self-contained and as elementary as possible. We have even made an extra effort to use things like the *double modulus* which one could argue that should be replaced by *ideals*. Although we do discuss ideals as part of the prologue to the actual discussion, we have kept the usage of anything abstract to a bare minimum and used them only if they were absolutely necessary. The thought of writing the whole book based on elementary methods only did cross our minds. However, writing everything from scratch and redeveloping known theory does not really favor the subject at hand in the long run. That being said, we used little abstract algebra most of which is only for the first two chapters. The rest of the chapters are almost free from abstract algebra concepts. Our purpose is to not only let an undergraduate student study these topics but also show an olympiad-level student how useful such theory can be in many cases. For this purpose, we have also added some exercises that appeared in various popular mathematical contests.

Contents

About the Authors

Masum Billal is working as a senior data scientist in Dhaka, Bangladesh. He has authored a book on number theory and he has been a trainer at math camps of Bangladesh for 5/6 years. His research interests include mathematics (specially number theory), machine learning, cryptography, etc.

Samin Riasat has completed his Master of Mathematics in Computer Science at the University of Waterloo, Canada. He has authored a book on mathematical olympiads and his research articles have been published in journals of repute. He is also a reviewer for *Mathematical Reviews*.

Chapter 1
Preliminaries

In this chapter, we discuss some topics from algebra that are prerequisites to the theory we will develop. First, we discuss groups, rings, fields, vector spaces, and matrices very briefly in Sects. 1.1 and 1.2. Then, we discuss properties of polynomials in Sect. 1.3 and cyclotomic polynomials in particular in Sect. 1.4. Later, we will talk about recurrent sequences (mostly linear recurrent sequences over fields) in Chap. 2. We will talk about divisibility sequences and explore their connections to binomial and related sequences in Chap. 3. We will use the theory developed in these chapters to characterize Lucas sequences in Chap. 4 and Lehmer sequences in Chap. 5. Finally, we will discuss special divisors of such sequences, particularly primitive divisors, in Chap. 6.

We will be using mostly elementary methods in this book. Occasionally we will require some advanced theory. However, our use of abstract theory will be minimal, to the point where we can even claim that it is still elementary. So a reader completely unfamiliar with abstract algebra should not feel overwhelmed . We could probably avoid using abstract theory altogether. For example, it is possible to only develop the theory of cyclotomic polynomials. However, doing so would make our discussion too restrictive and very case specific when we can obtain far more general results with the introduction of little theory. We initially wanted to avoid going into abstract theory altogether but soon realized that the amount of effort this will require will not be worth it.

To be more specific, we will need some basic ideas from linear and abstract algebra. In order to keep the book self-contained, we will discuss these topics from a beginner's perspective. We will be using a standard notation common to most texts for your convenience. You can consult texts like [Lan86, LN97] but we do not require such heavy machinery. Therefore, if you are unfamiliar with these subjects, you may find it easier to focus on what is presented in this book alone. We suggest you to use examples to make sense of the definitions and ideas. Then, if you feel like you want to dig deeper you can consult other texts. For example, linear algebra will be required mostly for the notion of a *basis*. And abstract algebra will be required

© The Author(s), under exclusive license to Springer Nature Singapore Pte Ltd. 2021
M. Billal and S. Riasat, *Integer Sequences*,
https://doi.org/10.1007/978-981-16-0570-3_1

for some properties of polynomials and matrices that would otherwise be difficult to describe. You may be wondering why we might need these things in a number theory text at all. To your surprise, you will see that recurrent sequences, polynomials, and matrices are very much connected to one another. This will make more sense once we develop the theories and show the connections. To give you an idea, most of the results regarding primitive divisors in Chap. 6 will be derived using cyclotomic polynomials.

1.1   Groups, Rings, Fields, and Vector Spaces

If you are completely new to this topic, do not be afraid. As long as you are familiar with sets, you should be able to grasp the ideas in this section. We also urge you to try to prove as many of the facts that we state as you can.

Vector spaces are just sets with some interesting properties. But before getting into vector space, we will discuss some more basic algebraic structures. These will help us define and develop a lot of properties with ease. First, we will talk about groups, rings, and fields. Then we will discuss vector spaces.

We mentioned that groups and vector spaces are just sets with interesting properties. One property that is common in all such algebraic structures is *closure*. We will be talking about binary operations defined on these sets. Such an operation takes two elements of the set as input. Now, it makes sense to be interested in the case when the element produced is again an element of the original set. This is why we call this property closure.

1.1.1   Groups

A *group* G is a set equipped with a binary operation $\circ$ satisfying the following properties. These properties are called *group axioms*.

(1) (Closure) For any $x, y \in G$, we have $x \circ y \in G$.
(2) (Associativity) For any $x, y, z \in G$,

$$x \circ (y \circ z) = (x \circ y) \circ z$$

(3) (Identity) There is an element $e \in G$ such that for any $x \in G$, we have $x \circ e = e \circ x = x$. This element e is called the *identity* of G.
(4) (Inverses) For any $x \in G$, there is an element $y \in G$ such that $x \circ y = e$. This element y is the *inverse* of x, usually denoted x^{-1} or $-x$.

Some consequences of these axioms are that the identity and the inverse of a group element must be unique, which justifies the above notation.

For example, the set $\mathbb{Z}$ of integers equipped with addition is a group. Observe that 0 is the identity of $\mathbb{Z}$, and $-x$ is the inverse of x for any $x \in \mathbb{Z}$. Note that the set $\mathbb{N}$ of natural numbers is not a group under addition. Also, $\mathbb{Z}$ is not a group under multiplication. The set $\{1, \omega, \omega^2\}$ where ω is an imaginary cubic root of unity ($\omega^3 = 1$) is a group under multiplication.

Note that a familiar property is missing from the list of properties of a group, which is commutativity. If

$$x \circ y = y \circ x$$

for any $x, y \in G$, then G is called an *abelian* group. It is however true in general that

$$(x \circ y)^{-1} = y^{-1} \circ x^{-1} \tag{1.1}$$

for any group G and $x, y \in G$. All the examples of groups we have mentioned so far are abelian. An example of a group that is not abelian is the set of invertible matrices under matrix multiplication. We will talk about this group near the end of Sect. 1.2.

Consider a group G with binary operation $\circ$. We write

$$x^n = \underbrace{x \circ \cdots \circ x}_{n \text{ times}}, \quad x^0 = e$$

for a positive integer n, and define $x^{-n} = (x^{-1})^n$. However, when $\circ$ represents addition $(+)$, it is more convenient to write x^n as nx and to use 0 for the identity element, so that $0x = 0$.

A subset of a group G that is also a group under the operation of G is a *subgroup* of G. The *order* of a group G is the number of elements $|G|$ of G. For an element $a \in G$, we write $\text{ord}(a)$ for the smallest positive integer k such that $a^k = e$. Observe that the set $\langle a \rangle = \{a^i : 0 \le i < |G|\}$ is a subgroup of G with $\text{ord}(a)$ elements. More generally, we define

$$\langle a_1, \ldots, a_n \rangle = \{a_1^{i_1} \cdots a_n^{i_n} : 0 \le i_j < |G|, 1 \le j \le n\}$$

If $G = \langle a_1, \ldots, a_n \rangle$, we call each a_i a *generator* of G. We call G *cyclic* if $G = \langle a \rangle$ for some a.

We give the following theorems without proof.

Theorem 1.1 *Every subgroup of a cyclic group is cyclic.*

Theorem 1.2 *Let $G = \langle a \rangle$ be a cyclic group of order n. The order of the subgroup $\langle a^k \rangle$ of G is $\frac{n}{\gcd(n,k)}$ for any positive integer k.*

Theorem 1.3 *Let $G = \langle a \rangle$ be a cyclic group of order n. If d is a divisor of n, then G has exactly $\varphi(d)$ elements of order d, where φ is the Euler Totient Function.*

Theorem 1.4 *Let $G = \langle a \rangle$ be a cyclic group of order n. Then $G = \langle a^i \rangle$ for each i with $1 \le i \le n$ and $\gcd(i, n) = 1$.*

The following theorem of Lagrange is a fundamental result in group theory and has a wide range of applications in all of mathematics. For instance, Fermat's little theorem and Euler's theorem are immediate consequences of Lagrange's theorem.

Theorem 1.5 (Lagrange's Theorem) *If H is a subgroup of G, then $|H|$ divides $|G|$.*

We will now generalize the notion of a group.

1.1.2 Rings and Fields

Consider a set S equipped with two operations $+$ (called addition) and $\cdot$ (called multiplication). We call S a *ring* if it satisfies the following properties.

(1) (Abelian Additive Group) S is an abelian group under $+$ with identity 0.
(2) (Associativity of Multiplication) For any $x, y, z \in S$,

$$x \cdot (y \cdot z) = (x \cdot y) \cdot z$$

(3) (Identity of Multiplication) There is an element $1 \in S$ such that $x \cdot 1 = 1 \cdot x = x$ for any $x \in S$. We call 1 the *multiplicative identity* of S.
(4) (Distributivity of Multiplication over Addition) For any $x, y, z \in S$,

$$x \cdot (y + z) = x \cdot y + x \cdot z$$
$$(y + z) \cdot x = y \cdot x + z \cdot x$$

Note that two familiar properties are missing for multiplication: commutativity and inverses. If S also has these properties, then S is called a *field*. That is, a field is a ring with the following additional properties.

i. (Commutativity of Multiplication) For any $x, y \in S$,

$$x \cdot y = y \cdot x$$

ii. (Inverses for Multiplication) For any $x \in S \setminus \{0\}$, there is an element $x^{-1} \in S$ such that $x \cdot x^{-1} = x^{-1} \cdot x = 1$.

If S is a field satisfying the properties above, we call S a field under the operations $+$ and $\cdot$. Notice that these are all familiar properties of real numbers. In fact, the set $\mathbb{R}$ of real numbers and the set $\mathbb{C}$ of complex numbers are both fields under their regular operations. The set $\mathbb{Q}$ of rational numbers likewise is also a field. However, the set $\mathbb{Z}$ of integers and the set $\mathbb{N}$ of natural numbers are not fields under their regular operations. If x is an element of $\mathbb{Z}$, then $-x$ is in $\mathbb{Z}$ but $\frac{1}{x}$ is not in $\mathbb{Z}$, since $\frac{1}{x}$ is not integer unless $|x| = 1$. And if x is an element of $\mathbb{N}$, then $-x$ is not always an element of $\mathbb{N}$.

As we can see, a field is simply a set with some desired properties. More generally, the objects studied in abstract algebra are sets with special properties. For instance, the set of all numbers of the form $a + b\sqrt{2}$, where a, b are rational numbers, is a field. Similarly, the set of numbers of the form $a + bi$, where $i = \sqrt{-1}$ and a, b are rational numbers, is also a field.

The non-zero elements of a field form an abelian group under multiplication. We write F^* for the set of non-zero elements of the field F and call F^* the *multiplicative group* of F. We also write xy instead of $x \cdot y$ for convenience.

Theorem 1.6 *If F is a field, then every finite subgroup of F^* is cyclic.*

If F, K are fields under the same operations with F a subset of K, we call F a *subfield* of K. If, in addition, $F \neq K$, then we call F a *proper subfield* of K. A field that has no proper subfield is called a *prime field*. Observe that $\mathbb{R}$ and $\mathbb{Q}$ are subfields of $\mathbb{C}$.

For a positive integer n, let $\mathbb{F}_n$ be the set $\{0, 1, 2, \ldots, n-1\}$ of residues modulo n. Then $\mathbb{F}_n$ is a ring under addition modulo n and multiplication modulo n, for any n. Moreover, $\mathbb{F}_p$ is a prime field for any prime p.

Let R be a ring. If there is a positive integer n such that $nx = 0$ for any $x \in R$, then the least such n is called the *characteristic* of the ring R. If no such n exists, then we say R has characteristic 0. For example, $\mathbb{F}_p$ has characteristic p, while $\mathbb{R}$ and $\mathbb{C}$ both have characteristic 0. A *domain* is a ring with no zero divisor; that is, if $xy = 0$ then either $x = 0$ or $y = 0$. If R is a domain with a non-zero characteristic, then the characteristic of R must be a prime number. As a result, all finite fields have a prime characteristic.

Theorem 1.7 *Let R be a commutative ring of prime characteristic p. Then for any $a, b \in R$ and integer $n \geq 0$,*

$$(a + b)^{p^n} = a^{p^n} + b^{p^n}$$

Proof We prove the result for $n = 1$ and leave the general case as an exercise. Recall the binomial theorem (see Binomial Theorem in the glossary),

$$(a + b)^n = a^n + \binom{n}{1}a^{n-1}b + \cdots + \binom{n}{n-1}a^1 b^{n-1} + b^n$$

Now, since $\binom{p}{k} = \frac{p}{k}\binom{p-1}{k-1}$, it follows that p divides $\binom{p}{k}$ for $0 < k < p$. Hence

$$(a + b)^p = a^p + \binom{p}{1}a^{p-1}b + \cdots + \binom{p}{1}ab^{p-1} + b^p$$

$$= a^p + b^p$$

as desired. ∎

Let F be a field, K a subfield of F and S a subset of F. Consider all the subfields of F that contain both K and S. The intersection of these subfields is also a field.

We call this field the *extension field of K adjoining S* and write it as $K(S)$. We call $K(S)$ an *extension* of K and if S is finite, a *finite extension* of K. Observe that $K(S)$ is the smallest subfield of F that contains both K and S. For a finite S, if $S = \{s_1, s_2, \ldots, s_k\}$, then we write $K(S) = K(s_1, s_2, \ldots, s_k)$. If $S = \{s\}$, then we call $K(s)$ a *simple extension*.

A set I is called an *ideal* of a ring R if $I \subseteq R$ and for all $x \in R, a \in I$, we have $xa \in I$. For instance, note that $\mathbb{Z}$ is not an ideal of $\mathbb{Q}$. For the ring $\mathbb{Z}$, the subset I which is the set of all multiples of some fixed integer n is an ideal of $\mathbb{Z}$. In fact, the smallest ideal containing a particular element $x \in R$ is the set $\{xa : a \in R\}$. We call this the *ideal generated by a*. An ideal I of R is called a *principal ideal* if is generated by some $a \in I$. We call I the principal ideal generated by a, and write $I = (a)$. A domain R is called a *principal ideal domain* if every ideal of R is a principal ideal. An ideal I of a ring R is called a *prime ideal* if $xy \in I$ only if $x \in I$ or $y \in I$ for all $x, y \in R$.

1.1.3   Vector Spaces

We now define a *vector space V* over a field F, which is a set with the following properties.

(1) (Abelian Additive Group) V is an abelian group under $+$ with identity 0.
(2) (Closure under Scalar Multiplication) If x is an element of V, then ax is an element of V for any a in F.
(3) (Properties of Scalar Multiplication) If a, b are elements of F and x, y are elements of V, then

$$a(x + y) = ax + ay$$
$$(a + b)x = ax + bx$$
$$a(bx) = (ab)x$$
$$0x = 0$$
$$1x = x$$

We call a subset U of V a *subspace* of V if it is a subgroup of V and is closed under scalar multiplication.

If F is a field, we denote by F^n the set of all n-tuples $(x_1, x_2, \ldots, x_n)$ where $x_1, x_2, \ldots, x_n$ are elements of F. Moreover, if $x = (x_1, x_2, \ldots, x_n)$ and $y = (y_1, y_2, \ldots, y_n)$, then we define

$$x + y = (x_1 + y_1, x_2 + y_2, \ldots, x_n + y_n)$$
$$ax = (ax_1, ax_2, \ldots, ax_n)$$

for a in F. Then F^n satisfies the properties of a vector space over F. In this case we just say F^n is a vector space and omit the mention of the base field. For example, $\mathbb{R}^n, \mathbb{C}^n, \mathbb{Q}^n$ are all vector spaces.

Theorem 1.8 *If F is an extension of K, then F is a vector space over K.*

If V is a vector space over a field F, then for $x_1, x_2, \ldots, x_n \in F$ and vectors $v_1, v_2, \ldots, v_n \in V$, we call $v_1 x_1 + v_2 x_2 + \cdots + v_n x_n$ a *linear combination* of $v_1, v_2, \ldots, v_n$.

Theorem 1.9 *For a vector space V over F and vectors $v_1, v_2, \ldots, v_n \in V$, the set of all linear combinations of $v_1, v_2, \ldots, v_n$ is a subspace of V.*

We call the subspace of V in Theorem 1.9 the *subspace generated by* $v_1, v_2, \ldots, v_n$.
Consider a vector space V over F and a set $B = \{v_1, v_2, \ldots, v_n\}$ of vectors in V. We say that B is *linearly dependent* if

$$a_1 v_1 + a_2 v_2 + \cdots + a_n v_n = 0$$

for some $a_1, a_2, \ldots, a_n \in F$ not all 0. If such elements do not exist, then we say that B is *linearly independent*. So, B is linearly independent if

$$a_1 v_1 + a_2 v_2 + \cdots + a_n v_n = 0$$

implies $a_1 = a_2 = \cdots = a_n = 0$. On the other hand, if every $v \in V$ can be written as a linear combination of $v_1, v_2, \ldots, v_n$, we say B *spans* V. If B is linearly independent and also spans V, then we call B a *basis* of V.

Theorem 1.10 *Let F be a field. The set of vectors*

$$v_1 = (1, 0, 0, \ldots, 0)$$
$$v_2 = (0, 1, 0, \ldots, 0)$$
$$\vdots$$
$$v_n = (0, 0, 0, \ldots, 1)$$

in F^n is a basis of F^n.

A subset B of V is called *maximally linearly independent* if B is linearly independent, but for any $v \in V \setminus B$, the set $B \cup \{v\}$ is linearly dependent. Likewise, a subset B of V is called *minimally spanning* if B spans V, but for any $b \in B$, the set $B \setminus \{b\}$ does not span V.

Theorem 1.11 *Let V be a vector space and $B \subseteq V$. Then the following are equivalent.*

1. *B is a basis of V.*
2. *B is maximally linearly independent.*
3. *B is minimally spanning.*

The *dimension* of a vector space V over a field F is the cardinality of a basis of V over F, and this is well-defined.

1.2 Matrices

Matrices are ubiquitous and an important object of study in mathematics. A matrix A over a field F is a rectangular array of elements from F.

We write

$$A = \begin{pmatrix} a_{1,1} & a_{1,2} & \cdots & a_{1,n} \\ a_{2,1} & a_{2,2} & \cdots & a_{2,n} \\ \vdots & \vdots & \ddots & \vdots \\ a_{m,1} & a_{m,2} & \cdots & a_{m,n} \end{pmatrix} \in F^{m \times n}$$

and call A an $m \times n$ *matrix over* F, where m and n are the number of rows and columns, respectively, and are called the *dimensions* of A. The entry in the i-th row and the j-th column is $a_{i,j}$. We can also use the shorthand $A = (a_{i,j})$, $1 \le i \le m$, $1 \le j \le n$ to represent A.

Observe that each row and each column of A represents a vector over F. For instance, we can express the i-th row of A as the $1 \times n$ matrix (or row vector)

$$A_{i,} = \begin{pmatrix} a_{i,1} & a_{i,2} & \cdots & a_{i,n} \end{pmatrix}$$

which represents $(a_{i,1}, a_{i,2}, \ldots, a_{i,n}) \in F^n$. Similarly, we can express the j-th column of A as the $m \times 1$ matrix (or column vector)

$$A_{,j} = \begin{pmatrix} a_{1,j} \\ a_{2,j} \\ \vdots \\ a_{m,j} \end{pmatrix}$$

which represents $(a_{1,j}, a_{2,j}, \ldots, a_{m,j}) \in F^m$.

We can also talk about operations on matrices. Naturally, two matrices of the same size can be added or subtracted coordinate-wise. That is, if $A, B \in F^{m \times n}$, we define $A \pm B = (a_{i,j} \pm b_{i,j})$. For example, if

$$A = \begin{pmatrix} 1 & 2 & 3 \\ 4 & 5 & 6 \\ 7 & 8 & 9 \end{pmatrix}$$

$$B = \begin{pmatrix} 10 & 11 & 12 \\ 13 & 14 & 15 \\ 16 & 17 & 18 \end{pmatrix}$$

then

$$A + B = \begin{pmatrix} 11 & 13 & 15 \\ 17 & 19 & 21 \\ 23 & 25 & 27 \end{pmatrix}$$

$$2A = A + A$$

$$= \begin{pmatrix} 2 & 4 & 6 \\ 8 & 10 & 12 \\ 14 & 16 & 18 \end{pmatrix}$$

$$B - 2A = \begin{pmatrix} 8 & 7 & 6 \\ 5 & 4 & 3 \\ 2 & 1 & 0 \end{pmatrix}$$

etc. Notice that this makes $F^{m \times n}$ into a group under $+$. The identity is the *zero matrix* 0, defined to be the matrix with all entries equal to 0. The (additive) inverse of A is $0 - A$, which we denote simply by $-A$. This also allows us to define equality of $A, B \in F^{m \times n}$ as

$$A = B \quad \text{if and only if} \quad A - B = 0 \tag{1.2}$$

In general, we say two matrices are equal if they have the same dimensions and satisfy 1.2.

For $c \in F$, we define $cA = (ca_{i,j})$. This defines scalar multiplication on $F^{n \times n}$ over F and allows $F^{m \times n}$ to be viewed as a vector space over F.

It is also possible in some sense to define multiplication and division for matrices. Traditional matrix multiplication is defined as follows. The product AB of two matrices A and B is well-defined if and only if the number of columns of A is equal to the number of rows of B. In particular, if A is $m \times n$ and B is $p \times q$, then AB is well-defined if and only if $n = p$. In this case, the product $C = AB = (c_{i,j})$ is the $m \times q$ matrix given by

$$c_{i,j} = \sum_{k=1}^{n} a_{i,k} b_{k,j}$$

It is important here to note that $AB \neq BA$ in general. In other words, matrix multiplication in general is not commutative.

For instance, with A and B in the example above,

$$AB = \begin{pmatrix} 1 \cdot 10 + 2 \cdot 13 + 3 \cdot 16 & 1 \cdot 11 + 2 \cdot 14 + 3 \cdot 17 & 1 \cdot 12 + 2 \cdot 15 + 3 \cdot 18 \\ 4 \cdot 10 + 5 \cdot 13 + 6 \cdot 16 & 4 \cdot 11 + 5 \cdot 14 + 6 \cdot 17 & 4 \cdot 12 + 5 \cdot 15 + 6 \cdot 18 \\ 7 \cdot 10 + 8 \cdot 13 + 7 \cdot 10 & 7 \cdot 11 + 8 \cdot 14 + 9 \cdot 17 & 7 \cdot 12 + 8 \cdot 15 + 9 \cdot 18 \end{pmatrix}$$

$$= \begin{pmatrix} 84 & 90 & 96 \\ 201 & 216 & 231 \\ 318 & 342 & 366 \end{pmatrix}$$

In order to understand the motivation behind this mechanism, we look at systems of linear equations. The simplest equations we deal with are usually linear equations of the form

$$2x = 6 \tag{1.3}$$

To solve (1.3), we divide both sides by 2 to obtain $x = 3$. Now, consider the system of linear equations

$$2x + 3y = 5$$
$$x + y = 2$$

It is routine to solve the system by hand. However, representing the system using matrices as follows can give us new insight.

$$\begin{pmatrix} 2 & 3 \\ 1 & 1 \end{pmatrix} \begin{pmatrix} x \\ y \end{pmatrix} = \begin{pmatrix} 2x + 3y \\ x + y \end{pmatrix}$$

So the system of equations can be represented using matrices as

$$\begin{pmatrix} 2x + 3y \\ x + y \end{pmatrix} = \begin{pmatrix} 5 \\ 2 \end{pmatrix}$$

$$\underbrace{\begin{pmatrix} 2 & 3 \\ 1 & 1 \end{pmatrix}}_{A} \underbrace{\begin{pmatrix} x \\ y \end{pmatrix}}_{X} = \underbrace{\begin{pmatrix} 5 \\ 2 \end{pmatrix}}_{B}$$

$$AX = B \tag{1.4}$$

We can see that (1.4) resembles (1.3). In order to solve (1.4) like (1.3) we wish to "divide" both sides of (1.4) by A. Note that this is equivalent to multiplying both sides of (1.4) *on the left* by a matrix A^{-1} that satisfies $A^{-1}A = I$. If successful, this would then give $X = A^{-1}B$, as desired. Such a matrix A^{-1} would be called the (left) inverse of A. Unfortunately, it does not always exist.

$$AX = B$$
$$(A^{-1})AX = A^{-1}B$$
$$(A^{-1}A)X = A^{-1}B$$
$$IX = A^{-1}B$$
$$X = A^{-1}B$$

Also note that it was important to multiply on the left because, as noted earlier, matrix multiplication is not commutative in general.

Ideally, we would like to view $F^{m \times n}$ as a ring. For this purpose it makes sense to consider the case $m = n$ in particular. We call an element of $F^{n \times n}$ a *square* matrix. So a matrix is square if it has an equal number of rows and columns. Let $A = (a_{i,j}) \in F^{n \times n}$ be a square matrix. A *diagonal matrix* is a square matrix with all off-diagonal entries equal to 0. Formally, A is *diagonal* if $a_{i,j} = 0$ whenever $i \neq j$. If A is diagonal with $a_{i,i} = 1$ for all i, then A is called the *identity* matrix and is denoted I_n, or simply I when the context is clear. So

$$I = \begin{pmatrix} 1 & 0 & 0 & \cdots & 0 \\ 0 & 1 & 0 & \cdots & 0 \\ 0 & 0 & 1 & \cdots & 0 \\ \vdots & \vdots & \vdots & \ddots & \vdots \\ 0 & 0 & 0 & \cdots & 1 \end{pmatrix}$$

With addition, multiplication, the zero matrix and the identity matrix defined as above, $F^{n \times n}$ becomes a ring.

It is natural to ask when the multiplication operation defined above is reversible. A square matrix A is said to be *invertible* or *non-singular* if there is a matrix A^{-1}, called the *inverse* of A, such that $A^{-1}A = AA^{-1} = I$. Otherwise we call A *singular*. An example of a singular matrix is

$$B = \begin{pmatrix} 1 & 2 \\ 2 & 4 \end{pmatrix}$$

Note that B corresponds, for example, to the following system of equations:

$$x + 2y = 4$$
$$2x + 4y = 9$$

which can be seen to have no solution, in accordance with the fact that B is a singular.

We give a few more useful definitions.

Let $A = (a_{i,j}) \in F^{m \times n}$. The *transpose* of A is defined to be the matrix $A^T = (a_{j,i})$. When $A = A^T$, we call A a *symmetric* matrix. When $m = n$, the *determinant* of A is defined recursively as $\det(A) = a_{1,1}$ if $n = 1$, and if $n > 1$, then

$$\det(A) = \sum_{j=1}^{n} (-1)^{j-1} a_{i,j} \det(M_{i,j})$$

for any i, where $M_{i,j}$ is the $(n-1) \times (n-1)$ matrix obtained from A by deleting its i-th row and j-th column. For example, if

$$A = \begin{pmatrix} a & b \\ c & d \end{pmatrix}$$

then $\det(A) = ad - bc$, and if

$$B = \begin{pmatrix} a & b & c \\ p & q & r \\ x & y & z \end{pmatrix}$$

then

$$\det(B) = a \det\left(\begin{pmatrix} q & r \\ y & z \end{pmatrix}\right) - b \det\left(\begin{pmatrix} p & r \\ x & z \end{pmatrix}\right) + c \det\left(\begin{pmatrix} p & q \\ x & y \end{pmatrix}\right)$$
$$= a(qz - ry) - b(pz - rx) + c(py - qx)$$

etc. Observe also that $\det(I) = 1$.

We now state the following theorems without proof.

Theorem 1.12 *If $A, B \in F^{n \times n}$, then $(AB)^T = B^T A^T$.*

Theorem 1.13 *If $A, B \in F^{n \times n}$, then $\det(AB) = \det(A) \det(B)$.*

Theorem 1.14 *A square matrix is invertible if and only if its determinant is non-zero.*

Theorem 1.15 *The determinant of a diagonal matrix is equal to the product of its diagonal entries. In particular, a diagonal matrix with non-zero diagonal entries is invertible.*

It follows from Theorems 1.13 and 1.14 that the product of two invertible matrices $A, B \in F^{n \times n}$ is again invertible, and by (1.1) that $(AB)^{-1} = B^{-1} A^{-1}$. In particular, the invertible matrices in $F^{n \times n}$ form a group under matrix multiplication, called the *general linear group* $\mathrm{GL}_n(F)$. It also follows from Theorem 1.13 that the matrices in $F^{n \times n}$ with determinant 1 form a group under matrix multiplication, called the *special linear group* $\mathrm{SL}_n(F)$. In fact, $\mathrm{SL}_n(F)$ is a subgroup of $\mathrm{GL}_n(F)$. Note that neither $\mathrm{GL}_n(F)$ nor $\mathrm{SL}_n(F)$ is abelian for $n > 1$ if F contains more than one element.

1.3 Polynomials

Let R be a ring. A *polynomial* f over R in the indeterminate x is an expression of the form

$$f(x) = a_k x^k + a_{k-1} x^{k-1} + \cdots + a_1 x + a_0$$

where $k \geq 0$ and, for each i, $a_i \in R$ is the *coefficient* of x^i in f. If $a_k \neq 0$, we call k the *degree* of f and write $\deg(f) = k$. If $\deg(f) > 0$, we call f *non-constant*, and call f *constant* otherwise. We say f is *monic* if $a_k = 1$. We denote the set of all polynomials over R as $R[x]$.

As before, we wish to define addition and multiplication for polynomials. Write $f(x) = \sum_{i=0}^{m} a_i x^i$ and $g(x) = \sum_{i=0}^{m} b_i x^i$ with $f, g \in R[x]$. Then we define $f + g$ and fg via

$$f(x) + g(x) = \sum_{i=0}^{m} (a_i + b_i) x^i$$

$$f(x)g(x) = \sum_{k=0}^{m+n} c_k x^k \quad \text{where} \quad c_k = \sum_{\substack{i+j=k \\ 0 \le i \le m \\ 0 \le j \le n}} a_i b_j$$

These make $R[x]$ into a ring with 0 and 1 coming from R. Observe that if $fg \ne 0$, then $\deg(f + g) \le \max\{\deg(f), \deg(g)\}$ and $\deg(fg) = \deg(f) + \deg(g)$.

We will mostly consider polynomials over fields. The reason being if F is a field, then $F[x]$ is a principal ideal domain. Consequently, we have a division algorithm for polynomials over a field. Henceforth, we write $a \mid b$ (resp. $a \nmid b$) to mean a *divides* (resp. *does not divide*) b.

Theorem 1.16 *Let F be a field and $f, g \in F[x]$ with g non-constant. Then $f = gq + r$ for unique polynomials $q, r \in F[x]$ with $r = 0$ or $\deg(r) < \deg(g)$.*

A non-constant polynomial $f \in F[x]$ is *reducible* if $f = gh$ for some non-constant polynomials $g, h \in F[x]$. Otherwise we call f *irreducible*. For example, any polynomial $f \in F[x]$ with $\deg(f) = 1$ is irreducible. On the other hand, $x^2 + 1$ is irreducible in $\mathbb{Z}[x]$ but is reducible in $\mathbb{C}[x]$ as $x^2 + 1 = (x + i)(x - i)$. Like prime numbers, if an irreducible polynomial $f \in F[x]$ divides a product $f_1 f_2 \cdots f_n$ with each $f_i \in F[x]$, then f divides f_i for some $1 \le i \le n$.

As a consequence of Theorem 1.16, we also obtain the following result.

Theorem 1.17 *If F is a field, then $F[x]$ is a principal ideal domain, and every ideal of $F[x]$ is of the form (f) for some monic polynomial $f \in F[x]$. In particular, every $f \in F[x]$ has a unique factorization of the form*

$$f = c f_1^{e_1} f_2^{e_2} \cdots f_m^{e_m}$$

where $c \in F$ and the f_i are distinct monic irreducible polynomials in $F[x]$.

Theorem 1.17 allows us to view polynomials over a field just like integers with unique factorization, enabling us to talk about their greatest common divisor (gcd).

Theorem 1.18 *Let $f \in \mathbb{F}_p[x]$. Then $f(x)^p = f(x^p)$.*

Proof Apply Theorem 1.7. ∎

An element $\alpha \in F$ is called a *root* (sometimes called a *zero*) of $f \in F[x]$ if $f(\alpha) = 0$. Here $f(\alpha)$ is calculated by substituting α for x in $f(x)$.

Theorem 1.19 (Remainder Theorem) *Let F be a field with $f \in F[x]$ and $\alpha \in F$. The remainder when f is divided by $x - \alpha$ is $f(\alpha)$.*

Proof By Theorem 1.16, there exist $q \in F[x]$ and $r \in F$ such that $f(x) = (x - \alpha)q(x) + r$. Substituting α for x gives $f(\alpha) = r$, as desired.   ∎

Theorem 1.20 (Factor Theorem) *Let F be a field with $f \in F[x]$ and $\alpha \in F$. Then $x - \alpha$ divides f if and only if $f(\alpha) = 0$.*

Proof This follows immediately from Theorem 1.19.   ∎

If α is a root of f, then the largest positive integer k for which $(x - \alpha)^k$ divides f is called the *multiplicity* of α. If $k > 1$, then we say that α is a *multiple root*. If f does not have a multiple root, then f is said to be *square-free*. Observe that a polynomial f in $F[x]$ can have at most $\deg(f)$ roots in F. Observe also that $x - \alpha$ for any $\alpha \in F$ is irreducible and square-free.

Let $f(x) = a_k x^k + \cdots + a_1 x + a_0 \in F[x]$. The *derivative* of f is the polynomial

$$f'(x) = ka_k x^{k-1} + \cdots + 2a_2 x + a_1 \in F[x]$$

Theorem 1.21 *Let F be a field and $f \in F[x]$. If $f = cf_1^{e_1} f_2^{e_2} \cdots f_m^{e_m}$ with $c \in F$ and the f_i distinct monic irreducible polynomials in $F[x]$, then*

$$f' = af_1^{e_1-1} f_2^{e_2-1} \cdots f_m^{e_m-1} \sum_{i=1}^{m} e_i f_1 f_2 \cdots f_{i-1} f_{i+1} \cdots f_m$$

Since $f_1^{e_1-1} f_2^{e_2-1} \cdots f_m^{e_m-1}$ divides both f and f', it follows immediately from Theorem 1.21 that α is a multiple root of f if and only if α is also a root of f'. In particular, we have the following.

Theorem 1.22 *Let F be a field and $f \in F[x]$. Then f is square-free if and only if $\gcd(f, f') = 1$.*

An immediate consequence is the following interesting result.

Theorem 1.23 *Let p be a prime number. If there is an integer α such that $x^n - 1 = (x - \alpha)^2 g(x)$ in $\mathbb{F}_p[x]$, then $p \mid n$. That is, $x^n - 1$ has a multiple root modulo p if and only if $p \mid n$.*

Proof Since $f(x) = x^n - 1 \in \mathbb{F}_p[x]$ is not square-free, it follows that $p \mid \gcd(x^n - 1, nx^{n-1})$. Therefore $p \mid n$.   ∎

For a polynomial $f \in F[x]$, the set $F[x]/(f)$ consists of the *residue classes* modulo f, written $g + (f)$ or $g \pmod{f}$ or $[g]$, for $g \in F[x]$. Two polynomials $g, h \in F[x]$ are said to be in the same residue class if and only if $g - h \in (f)$. In this case we write $g + (f) = h + (f)$ or $g \equiv h \pmod{f}$ or $[g] = [h]$.

We can define operations on the residue classes modulo f as follows.

$$[g] \pm [h] = [g \pm h]$$
$$[g][h] = [gh]$$

These operations are well-defined on $F[x]/(f)$ and make $F[x]/(f)$ into a ring, called the *residue class ring* or *quotient ring* of $F[x]$ by (f). The additive identity is $[0]$ and the multiplicative identity is $[1]$.

Theorem 1.24 *Let F be a field and $f \in F[x]$. The quotient ring $F[x]/(f)$ is a field if and only if f is irreducible over F.*

When $F = \mathbb{F}_p$ in Theorem 1.24, we can identify $F[x]/(f)$ with $\mathbb{F}_{p^d}$, where $d = \deg(f)$. So, $\mathbb{F}_{p^k}$ is a field for any prime p and positive integer k. These are in fact the only finite fields. Observe that $\mathbb{F}_{p^k}$ is an extension of $\mathbb{F}_p$.

If a polynomial f of degree n has roots $\alpha_1, \alpha_2, \ldots, \alpha_n$, then its *discriminant* is

$$\mathcal{D}(f) = (-1)^{\frac{n(n-1)}{2}} a_n^{2n-2} \prod_{1 \le i < j \le n} (\alpha_i - \alpha_j)$$

where a_n is the coefficient of x^n in $f(x)$. For example, the discriminant of the quadratic polynomial $ax^2 + bx + c$ is $b^2 - 4ac$. Observe that $\mathcal{D}(f) = 0$ if and only if f has a multiple root.

Note that $\mathcal{D}(f) = a_n^{2n-2} \det(V(f))$, where $V(f)$ is the *Vandermonde matrix* of f defined as

$$V(f) = \begin{pmatrix} 1 & \alpha_1 & \cdots & \alpha_1^{n-1} \\ 1 & \alpha_2 & \cdots & \alpha_2^{n-1} \\ \vdots & \vdots & \ddots & \vdots \\ 1 & \alpha_n & \cdots & \alpha_n^{n-1} \end{pmatrix}$$

We can similarly define the *resultant* or *eliminant* of two polynomials

$$f(x) = a_m x^m + \cdots + a_1 x + a_0$$
$$g(x) = b_n x^n + \cdots + b_1 x + b_0$$

as

$$\mathfrak{R}(f, g) = a_m^n b_n^m \prod_{i=1}^{m} \prod_{j=1}^{n} (\alpha_i - \beta_j)$$

where $\alpha_1, \alpha_2, \ldots, \alpha_m$ are the roots of f and $\beta_1, \beta_2, \ldots, \beta_n$ are the roots of g. Note that $\mathfrak{R}(f, g) = \det(S(f, g))$, where $S(f, g)$ is the *Sylvester matrix* of f, g defined as

$$S(f,g) = \left. \begin{pmatrix} a_m & a_{m-1} & \cdots & a_0 & & & \\ & a_m & a_{m-1} & \cdots & a_0 & & \\ & & \ddots & \ddots & \ddots & \ddots & \\ b_n & b_{n-1} & \cdots & b_0 & & & \\ & b_n & b_{n-1} & \cdots & b_0 & & \\ & & \ddots & \ddots & \ddots & \ddots & \end{pmatrix} \begin{array}{c} \left.\vphantom{\begin{matrix}a\\a\\a\end{matrix}}\right\}n \\ \\ \left.\vphantom{\begin{matrix}a\\a\\a\end{matrix}}\right\}m \\ \\ \end{array} $$

with the blank entries representing 0.

Let F be a field and K be a subfield of F. If $\alpha \in F$ is a root of a non-zero polynomial $f(x) \in K[x]$, then α is said to be *algebraic root* over K. If every $\alpha \in F$ is algebraic over K, then F is said to be an *algebraic extension* of K.

Let F be a field and K be a subfield of F. Let $\alpha \in F$ be algebraic over K. Observe that the set of polynomials $f \in K[x]$ for which $f(\alpha) = 0$ is an ideal of $K[x]$. By Theorem 1.17, this ideal is of the form (m) for some monic polynomial $m \in K[x]$. We call m the *minimal polynomial* of α over K. Observe that m is irreducible in $K[x]$. Furthermore, α is a root of some $f \in K[x]$ if and only if $m \mid f$. So, we have the following theorem.

Theorem 1.25 *Let F be a field and $f \in F[x]$ be an irreducible polynomial . If α is a root of f, then α is a root of $g \in F[x]$ if and only if $f \mid g$ in $F[x]$.*

If α is algebraic over $\mathbb{Q}$ with minimal polynomial m, we call α an *algebraic number*. If in addition $m \in \mathbb{Z}[x]$, we call α an *algebraic integer*. It is not hard to see that α is rational if and only if $\deg(m) = 1$.

Let K be a field and F be a finite extension of K. Then F is algebraic over K. If a polynomial $f \in K[x]$ has roots $\alpha_1, \alpha_2, \ldots, \alpha_n \in F$, then F is a *splitting field* of f.

Theorem 1.26 *Let p be a prime number and $f \in \mathbb{F}_p[x]$ be a polynomial of degree $d \geq 1$ such that $f(0) \neq 0$. Then there is a positive integer $k \leq p^d - 1$ such that $f \mid x^k - 1$.*

Proof The residue class ring $\mathbb{F}_p[x]/(f)$ contains $p^d - 1$ non-zero residue classes. Consider the elements $x^i + (f)$ for $0 \leq i \leq p^d - 1$ in $\mathbb{F}_p[x]/(f)$. Since $f(0) \neq 0$, it follows that $x \nmid f$, so $x^i \notin (f)$ for $0 \leq i \leq p^d - 1$. Hence there must be i, j with $0 \leq i < j \leq p^d - 1$ such that x^i and x^j belong to the same residue class. Then $x^i \equiv x^j \pmod{f}$, so $x^{j-i} \equiv 1 \pmod{f}$. Thus $f \mid x^{j-i} - 1$, as desired. $\blacksquare$

Theorem 1.26 enables us to define the *order* $\mathrm{ord}(f)$ of $f \in \mathbb{F}_p[x]$ to be the smallest positive integer d such that $f \mid x^d - 1$, which is well-defined if $f(0) \neq 0$. For an arbitrary f, write $f(x) = x^r g(x)$, where $\gcd(x, g(x)) = 1$. Then we can define $\mathrm{ord}(f)$ to be $\mathrm{ord}(g)$.

Consider an extension $\mathbb{F}_{p^n}$ of $\mathbb{F}_p$ and let $\alpha \in \mathbb{F}_{p^n}$. We call $\alpha, \alpha^p, \alpha^{p^2}, \ldots, \alpha^{p^{n-1}}$ the *conjugates* of α over $\mathbb{F}_p$. Observe that these are roots of the minimal polynomial m of α over $\mathbb{F}_p$. Moreover, they are distinct if and only if $\deg(m) \geq n$. Then we have the following.

Theorem 1.27 *The conjugates of* $\alpha \in \mathbb{F}_{p^n}^*$ *over* $\mathbb{F}_p$ *have the same order.*

Theorem 1.28 *Let* $f \in \mathbb{F}_p[x]$ *be an irreducible polynomial of degree* $d \geq 1$ *such that* $f(0) \neq 0$. *If* $\alpha \in \mathbb{F}_p^*$ *is a root of* f, *then* $\mathrm{ord}(f) = \mathrm{ord}(\alpha)$.

Proof Observe that $\mathbb{F}_{p^d}$ is a splitting field of f over $\mathbb{F}_p$. By Theorem 1.27, the roots of f have the same order in $\mathbb{F}_{p^n}^*$. Let $n = \mathrm{ord}(\alpha)$. Then α is a root of $x^n - 1$. Thus $f \mid x^n - 1$ and so $\mathrm{ord}(f) \leq n$. If $m = \mathrm{ord}(f) < n$ then $f \mid x^m - 1$, whence $\alpha^m = 1$, which is impossible. Thus $\mathrm{ord}(f) = n$. $\blacksquare$

Theorem 1.29 *Let* $f \in \mathbb{F}_p[x]$ *be an irreducible polynomial such that* $f(0) \neq 0$. *Then* $\mathrm{ord}(f) \mid p^{\deg(f)} - 1$.

Proof Note that $\mathbb{F}_{p^n}^*$ is a group of order $p^n - 1$. For α a root of f, if $\langle \alpha \rangle$ has order k, then $\mathrm{ord}(f) = \mathrm{ord}(\alpha) = k$ by Theorem 1.28. So, $k \mid p^n - 1$ by Theorem 1.5 . $\blacksquare$

Theorem 1.30 *Let* k *be a positive integer and* $f \in \mathbb{F}_p[x]$. *Then* $f \mid x^k - 1$ *if and only if* $\mathrm{ord}(f) \mid k$.

Proof If $f(0) = 0$ then we are done. Otherwise, let $d = \mathrm{ord}(f)$. Then $f \mid x^d - 1$.
If $d \mid k$, then $x^d - 1 \mid x^k - 1$. Hence $f \mid x^k - 1$.
Conversely, if $f \mid x^k - 1$, write $k = dq + r$ for integers q, r with $0 \leq r < d$.
Then

$$
\begin{aligned}
x^k - 1 &= x^{dq+r} - 1 \\
&= x^{dq} x^r - 1 \\
&= (x^{dq} - 1)x^r + x^r - 1 \\
&= (x^d - 1)g(x)x^r + x^r - 1
\end{aligned}
$$

for some polynomial $g \in \mathbb{F}_p[x]$. Since $f \mid x^k - 1$ and $f \mid x^d - 1$, we have $f \mid x^r - 1$. Since $r < d$, it follows that $r = 0$. Thus $d \mid k$. $\blacksquare$

The purpose of discussing all this background from abstract algebra was to introduce the next theorem. The theorems below will be used to develop the theory of linear recurrent sequences in Chap. 2.

Theorem 1.31 *Let* $f \in \mathbb{F}_p[x]$ *be an irreducible polynomial. Suppose that* $\mathrm{ord}(f) = d$ *and let* r *be the smallest positive integer such that* $p^r \geq k$. *Then* $\mathrm{ord}(f^k) = dp^r$.

Proof Since $f \mid x^d - 1$, it follows that $f^{p^r} \mid (x^d - 1)^{p^r}$. By Theorem 1.7,

$$
\begin{aligned}
(x^d - 1)^{p^r} &= (x^d)^{p^r} - 1 \\
&= x^{dp^r} - 1
\end{aligned}
$$

Thus $f^k \mid f^{p^r} \mid x^{dp^r} - 1$. Hence $\mathrm{ord}(f^k) \mid dp^r$ by Theorem 1.30.

On the other hand, since $f^k \mid x^{\mathrm{ord}(f^k)} - 1$, we have $f \mid x^{\mathrm{ord}(f^k)} - 1$. Therefore $d \mid \mathrm{ord}(f^k)$ by Theorem 1.30.

Hence $\mathrm{ord}(f^k) = dp^s$ for some $0 \leq s \leq r$.

By Theorem 1.23, $x^d - 1$ has no multiple root. Hence every root of $x^{dp^s} - 1 = (x^d - 1)^{p^s}$ has multiplicity p^s. Since $f^k \mid x^{dp^s} - 1 = (x^d - 1)^{p^s}$, it follows that $p^s \geq k$, whence $s \geq r$. Therefore $s = r$ and $\mathrm{ord}(f^k) = dp^r$. ∎

Theorem 1.32 *Let $f_1, f_2, \ldots, f_n \in \mathbb{F}_p[x]$ be irreducible pairwise relatively prime polynomials and put $f = f_1 f_2 \cdots f_n$. Then*

$$\mathrm{ord}(f) = \mathrm{lcm}(\mathrm{ord}(f_1), \mathrm{ord}(f_2), \ldots, \mathrm{ord}(f_n))$$

Using the theorems above we obtain the following general result.

Theorem 1.33 *Let $f \in \mathbb{F}_p[x]$ such that $f(0) \neq 0$. Suppose that $f = c f_1^{e_1} f_2^{e_2} \cdots f_k^{e_k}$, where $c \in \mathbb{F}_p$ and each f_i is a monic irreducible polynomial in $\mathbb{F}_p[x]$. Then*

$$\mathrm{ord}(f) = \mathrm{lcm}(\mathrm{ord}(f_1), \mathrm{ord}(f_2), \ldots, \mathrm{ord}(f_k)) p^r$$

where r is the smallest positive integer such that $p^r \geq \max(e_1, e_2, \ldots, e_k)$.

[War37, P. 279, Lemma 4.1] states the following two theorems. An *integer polynomial* here means a (possibly multi-variable) polynomial with integer coefficients. The first one is known as *Schatanovski's principle*.

Theorem 1.34 *Let $F(x_1, x_2, \ldots, x_k)$ be an integer polynomial symmetric in $x_1, x_2, \ldots, x_k$. Then for a positive integer m and an integer polynomial $f(x)$, if*

$$\begin{aligned} F(x) &\equiv (x - \alpha_1)(x - \alpha_2) \cdots (x - \alpha_k) \\ &\equiv (x - \beta_1)(x - \beta_2) \cdots (x - \beta_k) \quad (\bmod\ m) \end{aligned}$$

then

$$F(\alpha_1, \alpha_2, \ldots, \alpha_k) \equiv F(\beta_1, \beta_2, \ldots, \beta_k) \quad (\bmod\ m)$$

Theorem 1.35 *Let $f(x)$ and $F(x)$ be integer polynomials such that*

$$\begin{aligned} f(x) &= x^k - a_{k-1}x^{k-1} - \cdots - a_1 - a_0 \\ F(x) &= x^k - b_{k-1}x^{k-1} - \cdots - b_1 x - b_0 \end{aligned}$$

and $F(x)$ has roots $\alpha_1^n, \alpha_2^n, \ldots, \alpha_k^n$, where $\alpha_1, \alpha_2, \ldots, \alpha_k$ are the roots of $f(x)$. Then, for any positive integer $l \leq k$ and prime p, $p \mid \gcd(a_0, a_1, \ldots, a_{l-1})$ if and only if $p \mid \gcd(b_0, b_1, \ldots, b_{l-1})$.

Proof First we show that if $p \mid \gcd(a_0, a_1, \ldots, a_{l-1})$ then $p \mid \gcd(b_0, b_1, \ldots, b_{l-1})$. W rite

$$f(x) \equiv x^{k-l}(x^l - a_{l-1}x^{l-1} - \cdots - a_{k-1}) \quad (\bmod \ p)$$

Let $g(x) = x^{k-l}(x^l - a_l x^{l-1} - \cdots - a_{k-1})$ with roots $\gamma_1, \gamma_2, \ldots, \gamma_l$ and $\gamma_i = 0$ for $l < i \leq k$. Then

$$(x - \alpha_1)(x - \alpha_2) \cdots (x - \alpha_k) \equiv (x - \gamma_1)(x - \gamma_2) \cdots (x - \gamma_k) \quad (\bmod \ p)$$

Using Vieta's formulas (see Vieta's Formulas in the glossary), the coefficients of a polynomial can be expressed as symmetric polynomials in the roots of the polynomial. Let $\phi_i(f)$ be the elementary symmetric polynomial of degree i in the roots of f. By Theorem 1.34,

$$b_{k-i+1} \equiv \phi_i(f) \quad (\bmod \ p)$$
$$b_{k-i+1} \equiv \phi_i(g) \quad (\bmod \ p)$$

Now, consider the polynomial $G(x)$ with roots $\gamma_1^n, \gamma_2^n, \ldots, \gamma_k^n$. Again, using Vieta's Formulas, if c_i is the coefficient of x^i in $G(x)$, then

$$\phi_{k-i+1}(G) = \sum \gamma_1^n \gamma_2^n \cdots \gamma_{k-i+1}^n$$
$$= c_i$$

But if $i < l$, then at least one of the roots will be 0 in this product, hence $c_i \equiv 0$ $(\bmod \ p)$. Similarly, $b_i \equiv 0$ $(\bmod \ p)$ for $i < l$ and $p \mid b_i$ for $0 \leq i < l$.

We now prove that if $p \mid \gcd(b_0, b_1, \ldots, b_{l-1})$, then $p \mid \gcd(a_0, a_1, \ldots, a_{l-1})$. Note that

$$b_l \equiv \phi_{k-l+1}(G) \quad (\bmod \ p)$$
$$\equiv \sum \gamma_1^n \gamma_2^n \cdots \gamma_{k-l+1}^n \quad (\bmod \ p)$$
$$\equiv \gamma_1^n \gamma_2^n \cdots \gamma_{k-l+1}^n \quad (\bmod \ p)$$
$$\equiv (\gamma_1 \gamma_2 \cdots \gamma_{k-l+1})^n \quad (\bmod \ p)$$
$$\equiv c_l^n \quad (\bmod \ p)$$

Thus, if $p \nmid c_l$, then $p \nmid b_l$. Therefore, if $p \nmid b_i$ for $i < l$, then $p \nmid a_i$, as desired. ∎

1.4 Cyclotomic Polynomials

Cyclotomic polynomials have been extensively studied over the past few centuries. [Gau01, §§341, P. 599] proved that the cyclotomic polynomial $\Phi_p(x)$ is irreducible. [Sch46] and [Kro45] also proved the same result. [Kro54] and [Ded57] proved the general case that $\Phi_n(x)$ is irreducible. And mathematicians have kept working on this, for example, the method shown by [Syl79] was used by [War55] to prove that real Lehmer numbers have primitive divisors. However, in old times, notations were not as consistent and convenient as they are now. Different authors such as [Car29], [War55], [BV04], [Syl79] have used different notations throughout history. Ward followed in Carmichael's footprints, while other authors used whatever notation they deemed fit. Nowadays it is a different story. Even though sometimes $C_n(x)$ is used, most authors, e.g., [Lan12, Section 3, Chapter 6, Part 2, Roots of unity] use $\Phi_n(x)$. Cyclotomic polynomials have wide applications in number theory. However, most common elementary texts do not cover this topic in detail even though one can argue that the subject is elementary. We review cyclotomic polynomials briefly in this section for our purposes in this book.

It is known that every positive integer greater than 1 can be written as a product of primes. Here, primes are irreducible in the sense that they cannot be divided by any other positive integer greater than 1. Consider a similar factorization for polynomials. To be more specific, consider the factorization of $x^n - 1$ into irreducible polynomials. A general question is, how far can we keep factoring polynomials ? In the case of $x^n - 1$, simple algebra gives

$$x^2 - 1 = (x + 1)(x - 1)$$
$$x^3 - 1 = (x - 1)(x^2 + x + 1)$$
$$x^4 - 1 = (x - 1)(x + 1)(x^2 + 1)$$

Here we cannot further factorize any of the factors present into two or more non-constant polynomials with integer coefficients. Without such a restriction, however, one can obtain factorizations such as $x^2 + x + 1 = (x - \omega)(x - \omega^2)$, where ω is a complex cubic root of unity. The natural question that arises then is how we can find a general procedure of factoring $x^n - 1$ into irreducible polynomials. Fortunately, cyclotomic polynomials are exactly these irreducible polynomials. The complex factorization of $x^2 + x + 1$ should shed some light on how to obtain such a factorization in the general case.

We can define the *cyclotomic polynomial of order n* as

$$\Phi_n(x) = \prod_{\substack{1 \leq i \leq n \\ \gcd(i,n)=1}} (x - \zeta^i) \tag{1.5}$$

where ζ is a *primitive n-th root of unity*. The product is over all the positive integers less than n which are relatively prime to n. Recall that ζ is an *n-th root of unity* if $\zeta^n = 1$. At first, the definition (1.5) may seem intimidating. However, if you look at it closely and compare it with the cubic case, you will see that it is simply a generalization of that factorization.

The idea of factoring $x^n - 1$ with roots of unity is a lot more powerful than one would deem at first sight. It can even be connected with the history of Fermat's Last Theorem. In case you are not familiar with Fermat's Last Theorem , it states that

$x^n + y^n = z^n$ is not true for any positive integers x, y, z and $n > 2$. It has eluded mathematicians for over three centuries. Finally, Andrew Wiles and Richard Taylor proved it in around 1993–1994. Many prominent mathematicians failed to prove this theorem despite significant efforts. In the process, a lot of theories have been developed. Prior to the proof of the general case, mathematicians used to try to prove the theorem in special cases. [WE78] is a good read for a historical perspective on this. Euler proved the case $n = 3$. Fermat himself prove the case $n = 4$ using Pythagorean triplets. Legendre and Dirichlet proved the case $n = 5$ in 1823. Then Dirichlet proved the case $n = 14$ in 1832. Gabriel Lamé proved the case $n = 7$ using roots of unity in 1839. Note that $n = 7$ covers the case $n = 14$, but the converse is false. So, even Dirichlet failed to prove the case $n = 7$. Lamé believed that he could generalize the idea. The main idea behind the proof was to write

$$x^p + y^p = (x + y)(x + \zeta y)(x + \zeta^2 y) \cdots (x + \zeta^{p-1} y)$$

and if, in some sense, the factors can be shown to be pairwise relatively prime, then $x^p + y^p = z^p$ would imply that each factor above must be a p-th power. Then one could use an *infinite descent* argument. Lamé's work was based mostly on Liouville's. However, when Lamé presented this idea for the general case, Liouville himself was not very enthusiastic. This idea indeed seems to be too cheap for so many prominent mathematicians to miss for so long. For example, Lagrange showed the factorization above as well. Lamé forgot to take into account the complexities the introduction of complex numbers could present. So the concern was raised whether prime factorization would remain unique if complex numbers were introduced. Going forward, another mathematician Wantzel claimed he could prove unique factorization for $n \leq 4$. Then he said that it was easy to see the same argument applies for $n > 4$, which was wrong of course. One does not easily see that the same argument applies in general. Later, Kummer went on to introduce *ideal complex numbers* which, we can say, laid the foundations of modern algebraic number theory. Anyway, as you can see, the history of cyclotomic integers and polynomials is quite rich, dramatic, and influential. Let us now continue our original discussion.

The smallest positive integer k for which $\zeta^k = 1$ is called the *order* of ζ. We denote the order of ζ by ord(ζ). And ζ is a *primitive n-th root of unity* if ord(ζ) = n; that is, $\zeta^k = 1$ does not hold for any positive integer $k < n$. If ζ is a primitive n-th root of unity of order k, letting $n = kq + r$ with $0 \leq r < k$, we see that k must divide n. Otherwise, we would have $\zeta^r = 1$ with r smaller than k which would contradict the minimality of k.

Fix a primitive n-th root of unity ζ and consider the powers of ζ modulo n. We are interested in finding k such that ζ^k is a primitive n-th root of unity; that is, ord(ζ^k) = n. Letting ord(ζ^k) = d, we have $n \mid dk$, since ord(ζ) = n. If $g = \gcd(n, k)$ with $n = gm, k = gl$, and $\gcd(m, l) = 1$, then $m \mid d$. Using $\zeta^{dk} = 1$ along with $d = mr$ for some positive integer r, we get $\zeta^{nlr} = 1$. In order to make ord(ζ^k) minimal, we need to have $r = 1$. This gives ord(ζ^k) = $n / \gcd(n, k)$. If $\gcd(n, k) > 1$, then ord(ζ^k) = $n / \gcd(n, k) < n$. In this case, ζ^k is not primitive. Therefore, ζ^k is primitive only if $\gcd(k, n) = 1$. This shows that $\Phi_n(x)$ is a polynomial of degree $\varphi(n)$.

Primitive roots of unity are interesting and worth studying because they generate all roots of unity.

We will now formally prove that Φ_n defined in (1.5) is indeed a polynomial with integer coefficients, that is, a polynomial in $\mathbb{Q}[x]$. We first prove the following result.

Theorem 1.36 *For any positive integer n,*

$$x^n - 1 = \prod_{d|n} \Phi_d(x) \tag{1.6}$$

Proof It suffices to check that both sides have the same roots. The roots of the polynomial on the left are exactly the n-th roots of unity. The roots of the polynomial on the right are the primitive d-th roots of unity for each divisor d of n. Since each n-th root of unity must be a primitive dth root of unity for a unique divisor d of n and vice-versa, the conclusion follows.   ∎

This theorem also gives another proof of

$$n = \sum_{d|n} \varphi(d) \tag{1.7}$$

Applying the Möbius Inversion (see Möbius Function and Möbius Inversion) to (1.6) gives

$$\Phi_n(x) = \prod_{d|n} \left(x^{\frac{n}{d}} - 1 \right)^{\mu(d)} \tag{1.8}$$

Theorem 1.37 Φ_n *is monic in* $\mathbb{Q}[x]$.

The proof of this theorem relies on the fact that if $f, g \in \mathbb{Q}[x]$ are monic and $fg \in \mathbb{Q}[x]$, then $f, g \in \mathbb{Q}[x]$. Then we can simply induct on n to prove that $\Phi_n \in \mathbb{Q}[x]$.

Proof The claim is trivially true for $n = 1$. Assume that Φ_m is monic in $\mathbb{Q}[x]$ for all $m < n$. Then

$$x^n - 1 = \prod_{d|n} \Phi_d(x)$$

$$= \Phi_n(x) \prod_{\substack{d|n \\ d<n}} \Phi_d(x)$$

$$= \Phi_n(x) f(x)$$

where

$$f(x) = \prod_{\substack{d|n \\ d<n}} \Phi_d(x)$$

By the induction hypothesis, Φ_d is monic in $\mathbb{Q}[x]$ for $d < n$, so f is monic in $\mathbb{Q}[x]$. Hence Φ_n is the quotient of two monic polynomials in $\mathbb{Q}[x]$, so is monic in $\mathbb{Q}[x]$. Finally, since $x^n - 1 = \Phi_n(x)f(x) \in \mathbb{Q}[x]$, we conclude by the above fact that $\Phi_n \in \mathbb{Q}[x]$.   ∎

We now give a few results that can be used to compute Φ_n for different values of n.

Theorem 1.38 *For a positive integer n and a prime number p,*

$$\Phi_{np}(x) = \begin{cases} \Phi_n(x^p) & \text{if } p \mid n \\ \dfrac{\Phi_n(x^p)}{\Phi_n(x)} & \text{otherwise} \end{cases}$$

Proof Using (1.8),

$$\Phi_{np}(x) = \prod_{d \mid np} \left(x^{\frac{np}{d}} - 1 \right)^{\mu(d)}$$

$$= \prod_{d \mid n} \left(x^{\frac{np}{d}} - 1 \right)^{\mu(d)} \prod_{\substack{d \mid np \\ d \nmid n}} \left(x^{\frac{np}{d}} - 1 \right)^{\mu(d)}$$

$$= \prod_{d \mid n} \left((x^p)^{\frac{n}{d}} - 1 \right)^{\mu(d)} \prod_{\substack{d \mid np \\ d \nmid n}} \left(x^{\frac{np}{d}} - 1 \right)^{\mu(d)}$$

$$= \Phi_n(x^p) \prod_{\substack{d \mid np \\ d \nmid n}} \left(x^{\frac{np}{d}} - 1 \right)^{\mu(d)}$$

$$= \Phi_n(x^p)P(x)$$

where we split the product into two parts based on whether $d \mid n$.

First, consider the case $p \mid n$. To evaluate P, we need to determine when $d \mid np$ but $d \nmid n$. Let $n = p^\alpha m$ with $\alpha \geq 1$ and $p \nmid m$. Then $d \mid p^{\alpha+1}m$ but $d \nmid p^\alpha m$. Hence $p^2 \mid d$, so that $\mu(d) = 0$. This gives $P(x) = 1$.

Next, assume that $p \nmid n$. To evaluate P, we note that if $d \mid np$ but $d \nmid n$, then $d = ap$ with $a \mid n$ and $\gcd(a, p) = 1$. Then

$$\mu(d) = \mu(ap) = \mu(a)\mu(p) = -\mu(a)$$

using the multiplicative property of μ. Therefore,

$$P(x) = \prod_{a|n} \left(x^{\frac{np}{ap}} - 1 \right)^{\mu(ap)}$$

$$= \prod_{a|n} \left(x^{\frac{n}{a}} - 1 \right)^{-\mu(a)}$$

$$= \Phi_n(x)^{-1}$$

as desired.   ∎

This result extends to prime powers as follows.

Theorem 1.39 *Let p be a prime number and n, k be positive integers. Then*

$$\Phi_{p^k n}(x) = \begin{cases} \Phi_n(x^{p^k}) & \text{if } p \mid n \\ \dfrac{\Phi_n(x^{p^k})}{\Phi_n(x^{p^{k-1}})} & \text{otherwise} \end{cases}$$

Theorem 1.40 *If m, n are positive integers having the same prime factors such that $m \mid n$, then*

$$\Phi_n(x) = \Phi_m(x^{\frac{n}{m}})$$

Proof Let $\nu_p(n)$ denote the largest integer a such that $p^a \mid n$. Observe that $\nu_p(m) \geq 1$ for all $p \mid n$. In a similar fashion as before, if $d \mid n$ but $d \nmid m$, then $\nu_p(d) \geq \nu_p(m) + 1 \geq 2$ for some prime p. Then $\mu(d) = 0$. Assuming $n = mk$, we have

$$\Phi_n(x) = \prod_{d|n} \left(x^{\frac{n}{d}} - 1 \right)^{\mu(d)}$$

$$= \prod_{d|m} \left(x^{\frac{n}{d}} - 1 \right)^{\mu(d)} \prod_{\substack{d|n \\ d\nmid m}} \left(x^{\frac{n}{d}} - 1 \right)^{\mu(d)}$$

$$= \prod_{d|m} \left(x^{\frac{mk}{d}} - 1 \right)^{\mu(d)} \cdot 1$$

$$= \prod_{d|m} \left((x^k)^{\frac{m}{d}} - 1 \right)^{\mu(d)}$$

$$= \Phi_m(x^k)$$

$$= \Phi_m(x^{\frac{n}{m}})$$

as desired.   ∎

Theorem 1.41 *If $n > 1$ is odd, then $\Phi_{2n}(x) = \Phi_n(-x)$.*

Theorem 1.42 *If $n = p_1^{e_1} p_2^{e_2} \cdots p_k^{e_k}$ with the p_i distinct primes, then*

$$\Phi_n(x) = \Phi_{p_1 p_2 \cdots p_k}\left(x^{p_1^{e_1-1} p_2^{e_2-1} \cdots p_k^{e_k-1}}\right)$$

Theorem 1.43 *Let k, n be positive integers. Then $\Phi_k(x) \mid \Phi_k(x^n)$ if and only if* $\gcd(k, n) = 1$.

We now look at the prime divisors of cyclotomic polynomials evaluated at the positive integers. They allow us to establish some important results. Let a denote an arbitrary positive integer.

Theorem 1.44 *For a positive integer n and a prime divisor p of $\Phi_n(a)$, if $p \mid \Phi_d(a)$ for some $d \mid n$ with $d < n$, then $p \mid n$.*

Proof Since $a^n - 1 = \prod_{d \mid n} \Phi_d(a)$, if p is a prime divisor of $\Phi_d(a)$ and $\Phi_n(a)$ for some $d \mid n$ with $d < n$, then a is a multiple root of $\Phi_n(x)$ modulo p. Hence $p \mid n$ by Theorem 1.23. ∎

Theorem 1.45 *For a positive integer n and a prime divisor p of $\Phi_n(a)$, either $p \mid n$ or $p \equiv 1 \pmod{n}$.*

Proof If $p \mid \Phi_d(a)$ for some $d \mid n$ with $d < n$, then $p \mid n$ by Theorem 1.44. Otherwise, let $k = \text{ord}(a)$ in $\mathbb{F}_p^*$. We claim that $k = n$. This will imply $n \mid |\mathbb{F}_p^*| = p - 1$ by Theorem 1.5, and consequently, $p \equiv 1 \pmod{n}$.

To prove the claim, observe that $p \mid \Phi_n(a) \mid a^n - 1$, so that $k \mid n$. If $k < n$, then $p \mid a^k - 1$ implies $p \mid \Phi_d(a)$ for some $d \mid k$ by Theorem 1.36. Then $d \le k < n$, contradicting the minimality of n. ∎

An important special case is the following.

Theorem 1.46 *Every prime divisor of $1 + a + a^2 + \cdots + a^{p-1}$ is either p or congruent to 1 $\pmod{p}$.*

Proof Let q be a prime divisor of

$$
\begin{aligned}
1 + a + a^2 + \cdots + a^{p-1} &= \frac{a^p - 1}{a - 1} \\
&= \frac{\prod_{d \mid p} \Phi_d(a)}{a - 1} \\
&= \frac{\Phi_1(a)\Phi_p(a)}{\Phi_1(a)} \\
&= \Phi_p(a)
\end{aligned}
$$

By Theorem 1.45, either $q = p$ or $q \equiv 1 \pmod{p}$. ∎

Theorem 1.46 can be used to prove a special case of Dirichlet's theorem on arithmetic progressions, which states that if $\gcd(a, b) = 1$, then the sequence (x) with $x_n = an + b$ contains infinitely many prime numbers. Although not particularly relevant to our discussion, we present the result as a nice corollary to Theorem 1.46.

Theorem 1.47 *There are infinitely many primes congruent to 1 (mod n) for any positive integer n.*

Proof It suffices to consider the case $n > 1$. If the number of such primes is finite, let P be the product of all such primes and the prime divisors of n. Pick a positive integer k such that $\Phi_n(P^k) > 1$, so that it has at least one prime divisor q. Such a k must exist since $\Phi_n(x)$ is a non-constant monic polynomial. Now $q \mid P^{nk} - 1$, so $q \nmid P$. On the other hand, by Theorem 1.45, either $q \mid n$ or $q \equiv 1$ (mod n), hence $q \mid P$. This is a contradiction. $\blacksquare$

Theorem 1.48 *If m, n are positive integers such that $\gcd(\Phi_m(a), \Phi_n(a)) > 1$, then m/n is a prime power.*

Proof Let p be a prime divisor of $\gcd(\Phi_m(a), \Phi_n(a))$ and write $m = p^u b$ and $n = p^v c$ for integers $u, v \geq 0$ such that $p \nmid bc$. It suffices to prove that $b = c$.

Since $p \mid \Phi_m(a)$ and $p \mid \Phi_n(a)$, we have $p \mid a^m - 1$ and $p \mid a^n - 1$ by Theorem 1.36. Then $p \nmid a$. So, if $u = 0$, then $p \mid \Phi_b(a)$. Otherwise,

$$\Phi_m(a) = \frac{\Phi_b(a^{p^u})}{\Phi_b(a^{p^{u-1}})}$$

Since $p \mid \Phi_m(a)$, we have $p \mid \Phi_b(a^{p^u})$. By Fermat's little theorem, $a^{p^u} \equiv a$ (mod p) and so $\Phi_b(a^{p^u}) \equiv \Phi_b(a)$ (mod p). Thus, in every case, $p \mid \Phi_b(a)$ and similarly, $p \mid \Phi_c(a)$. Then $p \mid a^b - 1$ and $p \mid a^c - 1$ by Theorem 1.36, so that $p \mid a^h - 1$, where $h = \gcd(b, c)$. So by Theorem 1.36, $p \mid \Phi_d(a)$ for some $d \mid h$.

If $b \neq c$, we may assume that $b > c$. Then $h < b$. Hence $p \mid \Phi_d(a)$ and $p \mid \Phi_b(a)$ with $d \mid b$ and $d < b$. Therefore $p \mid b$ by Theorem 1.44. This contradicts $p \nmid bc$. Thus $b = c$. $\blacksquare$

The next theorem is from [Ser85, P. 133–134, Lemma 347]. It follows from Theorem 1.18.

Theorem 1.49 *Let p be a prime number and $s, m \geq 0$ be integers such that $p \nmid m$. Then*

$$\Phi_m(x^{p^s}) \equiv \Phi_m(x)^{p^s} \quad (\text{mod } p)$$

References

[BV04]   Geo. D. Birkhoff and H. S. Vandiver. "On the Integral Divisors of $a^n - b^n$". In: Ann. Mat. 5.4 (1904), pp. 173-180. issn: 0003486X. http://www.jstor.org/stable/2007263

[Car29] R. D. Carmichael. "A Simple Principle of Unification in the Elementary Theory of Numbers". In: Amer. Math. Monthly 36.3 (1929), pp. 132-143. issn: 00029890, 19300972. http://www.jstor.org/stable/2299678

[Ded57] R. Dedekind, Beweis für die Irreductibilität der Kreistheilungs-Gleichungen. Journal für die reine und angewandte Mathematik **54**, 27–30 (1857)

[Gau01] Carl Friedrich Gauss. Disquisitiones arithmeticae. 1801

[Kro45] L. Kronecker, Beweis dass für jede Primzahl p die Gleichung $1 + x + x^2 + +x^{p-1} = 0$ irreductibel ist. Journal für die reine und angewandte Mathematik (Crelles Journal) **1845**(29), 280–280 (1845). https://doi.org/10.1515/crll.1845.29

[Kro54] L. Kronecker, Memoire sur les facteures irréductibles de l'expression $x^n - 1$. Journal de mathématiques pures et appliquées : ou recueil mensuel de mémoires sur les diverses parties des mathématiques **19**, 177–192 (1854)

[Lan86] Serge Lang, *Introduction to Linear Algebra* (Springer, New York, 1986)

[Lan12] Serge Lang. Algebra. Springer-verlag New York Inc., 2012

[LN97] Rudolf Lidl and Harald Niederreiter. Finite Fields: Encyclopedia of Mathematics and Its Applications. Vol. 7. 1997, p. 136

[Sch46] L. Schönemann. "Von denjenigen Moduln, welche Potenzen von Primzahlen sind." In: Journal für die reine und angewandte Mathematik 32 (1846), pp. 93-105. http://eudml.org/doc/147334

[Ser85] Joseph Alfred Serret. Cours d'Algèbre supérieure Tome 1. fre. Paris: Gauthier-Villars, 1885. <error l="292" c="Undefined command " />http://eudml.org/doc/202932

[Syl79] J. J. Sylvester. "On Certain Ternary Cubic-Form Equations". In: Amer. J. Math. 2.4 (1879), pp. 357-393. issn: 00029327, 10806377. http://www.jstor.org/stable/2369490

[War37] Morgan Ward. "Linear Divisibility Sequences". In: Trans. Amer. Math. Soc. 41.2 (1937), pp. 276-286. issn: 00029947. http://www.jstor.org/stable/1989623

[War55] Morgan Ward. "The Intrinsic Divisors of Lehmer Numbers". In: Ann. Mat. 62.2 (1955), pp. 230-236. issn: 0003486X. http://www.jstor.org/stable/1969677

[WE78] Lawrence Washington and Harold M. Edwards. "Fermat's Last Theorem: A Genetic Introduction to Algebraic Number Theory." In: Mathematics of Computation 32.143 (1978), p. 943. doi: https://doi.org/10.2307/2006502.

Chapter 2
Linear Recurrent Sequences

In this chapter, we discuss linear recurrent sequences over a field. We give results on when such a sequence is periodic and obtain an upper bound on the length of the period, as well as show how to produce this bound. Finally, we discuss the theory developed by Morgan Ward on the periodicity of such sequences with the help of the *double modulus*. We will see a lot of results that are of fundamental importance in this theory.

2.1 Introduction

A sequence (a) of integers defined by

$$a_{n+k} = c_{k-1}a_{n+k-1} + \cdots + c_1 a_{n+1} + c_0 a_n + \alpha \qquad (2.1)$$

where a_i is the i-th term of the sequence and the c_i are constants, is called a *(linear) recurrent sequence* of *order* k. Note that like monic polynomials, $c_k = 1$. We call (a) homogeneous if $\alpha = 0$, otherwise call (a) *non-homogeneous*. The order here is k and not $k + 1$ because a_{n+k} is represented as a linear combination of the k preceding terms of the sequence. This will make more sense when we define the *characteristic polynomial* of (a) below. Recurrent sequences have also been called *solutions to linear difference equations* [Bel30, War33a].

We will mostly focus on homogeneous sequences. In this chapter we discuss some properties of recurrent sequences, especially periodicity. Considering (2.1) modulo a positive integer m, we get a new sequence (r) of residues where r_i is the least non-negative residue of a_i upon division by m. We call (r) the *reduction* of (a) modulo m.

Consider the recurrent sequence (a) defined in (2.1) over the finite field $\mathbb{F}_m$, so that the coefficients $c_{k-1}, \ldots, c_0$ and the terms $a_{n+k-1}, \ldots, a_n$ are elements of

© The Author(s), under exclusive license to Springer Nature Singapore Pte Ltd. 2021
M. Billal and S. Riasat, *Integer Sequences*,
https://doi.org/10.1007/978-981-16-0570-3_2

$\mathbb{F}_m$. Here we assume that $m = p^s$ for a prime p and a positive integer s. We call $s_i = (a_i, a_{i+1}, \ldots, a_{i+k-1})$ the i-th *state vector* of (a). State vectors will be very useful for characterizing periodicity of recurrent sequences as we will see later.

We now define the *characteristic polynomial* of a recurrent sequence. The motivation behind this comes from assuming that the terms of a recurrent sequence (a) can be written as the powers of some constant c; that is, $a_n = c^n$ for all n, so that (a) is in fact just a geometric sequence. This may seem like a wild guess, but it is not completely unreasonable. For example, the recurrent sequence $a_n = 2a_{n-1}$ can be obtained this way. Another non-trivial example is the Fibonacci Sequence (F) given by

$$F_{n+1} = F_n + F_{n-1} \tag{2.2}$$

Although this is not exactly a geometric sequence, it has an interesting property. Looking at the ratio of consecutive Fibonacci numbers we observe that

$$\lim_{n \to \inf} \frac{F_{n+1}}{F_n}$$

exists. This suggests that (F) behaves in some sense like a geometric sequence. To find the *common ratio*, we let

$$\phi = \lim_{n \to \infty} \frac{F_{n+1}}{F_n}$$
$$= \lim_{n \to \infty} 1 + \frac{F_{n-1}}{F_n}$$
$$= 1 + \frac{1}{\phi}$$
$$\phi^2 = 1 + \phi \tag{2.3}$$
$$\phi = \frac{1 \pm \sqrt{5}}{2}$$

As a matter of fact, we have

$$F_n = \frac{1}{\sqrt{5}} \left(\left(\frac{1 + \sqrt{5}}{2} \right)^n - \left(\frac{1 - \sqrt{5}}{2} \right)^n \right)$$

So F_n is a linear combination of the powers of the roots of (2.3). To understand this behavior, say it possible to write $F_n = r^n$ by *wishful thinking*. Then $r^n = r^{n-1} + r^{n-2}$, so that $r^2 = r + 1$, which is exactly (2.3). If the roots of this equation are r_1 and r_2, then this means that $F_n = r_1^n$ and $F_n = r_2^n$ satisfy (2.2). Just like how the set of solutions to a linear system of equations is a vector space, it turns out that the set of solutions to a linear recurrent sequence is also a vector space. In particular, the general solution to (2.2) is of the form

$$F_n = b_1 r_1^n + b_2 r_2^n$$

for some constants b_1, b_2. We can solve for b_1, b_2 using the initial values $F_0 = 0$, $F_1 = 1$ of the Fibonacci Sequence and obtain

$$F_n = \frac{r_1^n - r_2^n}{r_1 - r_2}$$

The above analysis worked because the roots of the characteristic polynomial of (F) were distinct. If they were not distinct, then we would expect the solution to be

$$\lim_{r_1 \to r_2} \frac{r_1^n - r_2^n}{r_1 - r_2} = n r_2^{n-1}$$

We can extend this idea to the general sequence (2.1). Assuming $\alpha = 0$, so that the sequence (a) is homogeneous, the *characteristic polynomial* of (a) is

$$f(x) = x^k - c_{k-1} x^{k-1} - \cdots - c_1 x - c_0$$

Theorem 2.1 *Given the initial state vector* $(a_0, a_1, \ldots, a_{k-1})$ *of the homogeneous recurrent sequence* (a) *of order k, if the characteristic polynomial* $f(x)$ *of* (a) *has k distinct roots* $\alpha_1, \alpha_2, \ldots, \alpha_k$, *then for any* $n \in \mathbb{N}$,

$$a_n = b_1 \alpha_1^n + b_2 \alpha_2^n + \cdots + b_k \alpha_k^n$$

for some constants $b_1, b_2, \ldots, b_k$ *that are uniquely determined by the initial state vector.*

As in the case of the Fibonacci Sequence, the assumption that the roots of $f(x)$ are distinct is very important here. We will simply mention when this theorem is applicable.

The characteristic polynomial $f(x)$ of (2.1) can be obtained in another way. Consider the $k \times k$ matrix C defined as follows. If $k > 1$, then

$$C = \begin{pmatrix} 0 & 0 & \cdots & 0 & c_0 \\ 1 & 0 & \cdots & 0 & c_1 \\ 0 & 1 & \cdots & 0 & c_2 \\ \vdots & \vdots & \ddots & \vdots & \vdots \\ 0 & 0 & \cdots & 1 & c_{k-1} \end{pmatrix}$$

and otherwise, $C = (c_0)$. The matrix C is called the *companion matrix* of (a). As we will soon see, this matrix will play a big role in our theory. For now, just notice that the characteristic polynomial can be written as

$$f(x) = \det(xI - C)$$

A matrix M is called *diagonalizable* if there exists an invertible matrix S such that $S^{-1}MS$ is a diagonal matrix . It turns out that C is diagonalizable if the roots of $f(x)$ are distinct. For example, the companion matrix of $a_n = 2a_{n-1} - a_{n-2}$ is

$$C = \begin{pmatrix} 0 & -1 \\ 1 & 2 \end{pmatrix}$$

which cannot be diagonalized.

A very important property of the companion matrix is its association with the state vectors. One can easily check by induction that

$$s_n = s_0 C^n$$

Theorem 2.1 can be generalized to non-homogeneous recurrences as follows. For a non-homogeneous linear recurrence (a) of order k given by (2.1), if the characteristic polynomial $f(x)$ of (a) has k distinct roots $\alpha_1, \alpha_2, \ldots, \alpha_k$, then

$$a_n = \frac{\alpha}{f(1)} + b_1 \alpha_1^n + b_2 \alpha_2^n + \cdots + b_k \alpha_k^n$$

As in Theorem 2.1, we obtain the unique constants $b_1, b_2, \ldots, b_k$ from the initial state vector $(a_0, a_1, \ldots, a_{k-1})$.

2.2 Periodicity

Consider a sequence of integers (henceforth referred to as an *integer sequence*) (a). We are interested in the *periodicity* of this sequence. It is possible to interpret periodicity in multiple ways. As usual, when we say a sequence is periodic, we mean the sequence repeats from some point. In particular, if there exist positive integers n_0, ρ such that $a_{n+\rho} = a_n$ for all $n \geq n_0$, then we say (a) is *ultimately periodic*. If $n_0 = 0$, then we simply call (a) *periodic*. Here we call n_0 the *pre-period* and ρ the *period*. Note, however, that these values may not be unique; we are mostly interested their smallest values. Carmichael [Car29] and Ward called this smallest ρ the *characteristic* of (a). However, this term can be easily confused with other things termed *characteristic* so we will simply call it the *least period* (or just *period* when there is no ambiguity) of (a).

For example, if (r) is the reduction of the recurrent sequence (a) modulo m, then we will show later that (r) is ultimately periodic. In particular, if every term of (a) is divisible by m, then we call (a) a *null sequence* modulo m.

Recall our definition of $\mathrm{ord}(f)$ for a polynomial f. If $f(x) = x^h g(x)$ for some $h \geq 0$ such that $\gcd(g(x), x) = 1$, equivalently $g(0) \neq 0$, then we defined $\mathrm{ord}(f)$ as the smallest positive integer d for which $g(x) \mid x^d - 1$. This was also a pre-condition for some theorems in Chapter 1. This will now make better sense as follows. If the characteristic polynomial $f(x)$ of a linear recurrent sequence (a) is of the form $x^h g(x)$, where $\gcd(g(x), x) = 1$, then the first h terms of the initial state vector s_0 can be ignored in the sense that they will have no impact on the periodicity of (a). Thus, we have the following theorem.

Theorem 2.2 *Let (a) be a linear recurrent sequence of order k with characteristic polynomial $f(x) = x^h g(x)$, where $\gcd(g(x), x) = 1$. Then the sequence (c) with $c_i = a_{i+h}$ has characteristic polynomial $g(x)$.*

We first give some results on the period of a recurrent sequence.

Theorem 2.3 *Let m be a positive integer and (a) be a recurrent sequence of order k over $\mathbb{F}_m$. Then (a) is ultimately periodic with period*

$$\rho \leq \begin{cases} m^k & \text{if } (a) \text{ is non-homogeneous} \\ m^k - 1 & \text{otherwise} \end{cases}$$

Proof The number of different possible state vectors of (a) is m^k. So, among the state vectors s_i for $0 \leq i \leq m^k$, there must be two that are equal. In particular, there exists $0 \leq i < j \leq m^k$ such that $s_i = s_j$. Incorporating this into (2.1) gives $s_{n+j-i} = s_n$ for all $n \geq i$. Hence (a) is ultimately periodic with period $\rho = j - i \leq m^k$. If (a) is homogeneous, then excluding the zero vector from our count gives $\rho \leq m^k - 1$. ∎

So any recurrent sequence over a finite field is ultimately periodic, although not necessarily periodic . It is easy to construct an ultimately periodic sequence that is not periodic.

Theorem 2.4 *Let F be a finite field and (a) be a recurrent sequence over F given by*

$$a_{n+k} = c_{k-1}a_{n+k-1} + \cdots + c_1 a_{n+1} + c_0 a_n + \alpha$$

If $c_0 \neq 0$, then (a) is periodic.

Proof We want to show that (a) has pre-period 0. Let ρ and n_0 be the least period and pre-period of (a) respectively. Then $a_{n+\rho} = a_n$ for all $n \geq n_0$. Since $c_0 \neq 0$, there is an inverse $c_0^{-1} \in F$. Then if $a_{n+\rho} = a_n$ for some $n \geq n_0$, we can obtain $a_{n+\rho-1} = a_{n-1}$ as follows. If $k \geq 1$, then $a_{n+k+\rho-1} = a_{n+k-1}$, so

$$a_{n+k+\rho-1} = c_{k-1}a_{n+k+\rho-1} + \cdots + c_0 a_{n+\rho-1} + \alpha$$
$$a_{n+\rho-1} = c_0^{-1}\left(a_{n+k+\rho-1} - c_{k-1}a_{n+\rho-2+k} - \cdots - c_1 a_{n+\rho} - \alpha\right)$$
$$= c_0^{-1}\left(a_{n+k-1} - c_{k-1}a_{n-2+k} - \cdots - c_1 a_n - \alpha\right)$$

and

$$a_{n+k-1} = c_{k-1}a_{n+k-2} + \cdots + c_0 a_{n-1} + \alpha$$

$$a_{n-1} = c_0^{-1}(a_{n+k-1} - c_{k-1}a_{n-2+k} - \cdots - c_1 a_n - \alpha)$$

Thus, $a_{n+\rho-1} = a_{n-1}$ and by induction, $a_{n+\rho-i} = a_{n-i}$ for $0 \le i \le n$. This shows that $n_0 = 0$ and $a_{n+\rho} = a_n$ for all $n \ge 0$, as desired. ∎

Theorem 2.5 *Let F be a finite field. If ρ is the least period of the recurrent sequence (a) over F and d is a period of (a), then $\rho \mid d$.*

Proof By definition, we have $a_{n+\rho} = a_n$ and $a_{n+d} = a_n$ for all n. Let $d = q\rho + r$ with $0 \le r < \rho$.

$$
\begin{aligned}
a_n &= a_{n+d} \\
&= a_{n+q\rho+r} \\
&= a_{n+(q-1)\rho+r} \\
&= \cdots \\
&= a_{n+r}
\end{aligned}
$$

If $r > 0$, then we have a period r with $0 < r < \rho$, which is a contradiction. Thus $\rho \mid d$. ∎

Bell [Bel30] gave the next theorem that connects the characteristic polynomial to the least period of a recurrent sequence. This is a very nice result and while we could use the theory we developed for fields, we present his original proof that uses only the ideas developed in this section.

Theorem 2.6 *Let (a) be a recurrent sequence of order k with an irreducible characteristic polynomial f. Then (a) has a unique least period ρ if and only if $k = \varphi(\rho)$ and the roots of f are the primitive ρ-th roots of unity.*

Proof Let $\alpha_1, \alpha_2, \ldots, \alpha_k$ be the roots of f. Since f is irreducible, the roots are distinct and $\mathcal{D}(f) \ne 0$. Then we have

$$a_n = b_k \alpha_k^n + \cdots + b_1 \alpha_1^n$$

for some $b_1, b_2, \ldots, b_k$ are some constant real numbers. These constants $b_1, b_2, \ldots, b_k$ are uniquely determined by the initial state vector $(a_0, a_1, \ldots, a_{k-1})$.

Since $a_{n+\rho} = a_n$ for all n, we have

$$b_k \alpha_k^{n+\rho} + \cdots + b_1 \alpha_1^{n+\rho} = b_k \alpha_k^n + \cdots + b_1 \alpha_1^n$$
$$b_k \alpha_k^n (\alpha_k^\rho - 1) + \cdots + b_1 \alpha_1^n (\alpha_1^\rho - 1) = 0$$

Observe that none of the roots are 0. Setting $n = 0, 1, \ldots, k-1$ we obtain the following system of equations.

$$b_k(\alpha_k^\rho - 1) + \cdots + b_1(\alpha_1^\rho - 1) = 0$$
$$b_k\alpha_k(\alpha_k^\rho - 1) + \cdots + b_1\alpha_1(\alpha_1^\rho - 1) = 0$$

$$\vdots$$

$$b_k\alpha_k^{k-1}(\alpha_k^\rho - 1) + \cdots + b_1\alpha_1^{k-1}(\alpha_1^\rho - 1) = 0$$

We can see that the system has a solution only if $b_i(\alpha_k^\rho - 1) = 0$ for $1 \le i \le k$. So, first we show that $b_i \ne 0$ for $1 \le i \le k$. Let h be the number of i such that $b_i = 0$. Then (a) can be written as a recurrent sequence of order $r - h$. If the characteristic polynomial in this case is g, then g shares those $r - h$ roots with f. So g is an integer polynomial that divides f. This forces g to be constant, whence $h = 0$. Thus, $\alpha_i^\rho - 1 = 0$ and α_i is a ρ-th root of unity. Furthermore, α_i must be a primitive ρ-th root of unity as ρ is the smallest positive integer that satisfies $\alpha_i^\rho = 1$. Finally, the number of primitive ρ-th roots is $\varphi(\rho)$. So, we must have $k = \varphi(\rho)$. ∎

So we can express the characteristic polynomial $f(x)$ as

$$\prod_{d|\rho}(x^d - 1)^{\mu\left(\frac{n}{d}\right)}$$

which is the cyclotomic polynomial $\Phi_\rho(x)$. [War33a] extended this theorem as follows.

Theorem 2.7 *Let (u) be a recurrent sequence of order k such that for fixed positive integers a, b we have*

$$u_a = u_{a+b} = u_{a+2b} = \cdots = u_{a+kb} \ne 0 \tag{2.4}$$

If the characteristic polynomial f of (u) is irreducible in $\mathbb{Z}[x]$, then (u) is periodic and the roots of f are roots of unity.

Proof Let $\alpha_1, \alpha_2, \ldots, \alpha_k$ be the roots of f. Since f is irreducible, the roots are distinct and we can write u_n as

$$u_n = b_1\alpha_1^n + b_2\alpha_2^n + \cdots + b_k\alpha_k^n$$

for some non-zero constants b_i. Let $c = u_{a+ib}$ for $0 \le i \le k$. We get the following system of equations from (2.4)

$$(b_1\alpha_1)\alpha_1^{ib} + (b_2\alpha_2)\alpha_2^{ib} + \cdots + (b_k\alpha_k)\alpha_k^{ib} - c = 0$$

for $0 \le i \le k$. By assumption, $c \ne 0$ and we have $k + 1$ homogeneous linear equations in $b_1\alpha_1, b_2\alpha_2, \ldots, b_k\alpha_k$, and c. So the determinant of the coefficient matrix of the system must be 0. We can write this determinant as the Vandermonde determinant $V(\alpha_1^b, \alpha_2^b, \ldots, \alpha_k^b, 1)$.

$$V(\alpha_1^b, \alpha_2^b, \ldots, \alpha_k^b, 1) = 0$$

$$\prod_{i \neq j}(\alpha_i^b - \alpha_j^b)\prod(\alpha_i^b - 1) = 0$$

Hence $\alpha_i^b - \alpha_j^b = 0$ or $\alpha_i^b - 1 = 0$ for some i. If $\alpha_i^b - 1 = 0$, every root of f must be a root of unity since f is irreducible. Otherwise, group the b-th powers of α_i into d sets of equal powers where $k = dl$.

$$\alpha_{il+1}^b = \alpha_{il+2}^b = \cdots = \alpha_{il+l}^b = \beta_{i+1}$$

Here, we have a grouping for each $0 \leq i < d$ and $\beta_1, \beta_2, \ldots, \beta_d$ are distinct irrational algebraic numbers. If $d = 1$, then β_1 is an integer and we see that in fact $\beta_i = \pm 1$. However, if $d > 1$, then for $0 \leq i \leq d$,

$$(b_1\alpha_1^a + b_2\alpha_2^a + \cdots + b_l\alpha_l^a)\beta_1^i + (b_{l+1}\alpha_{l+1} + \cdots + b_{2l}\alpha_{2l})\beta_2^i + \cdots - c = 0$$

As before, we see that

$$V(\beta_1, \beta_2, \ldots, \beta_d, 1) = 0$$

$$\prod_{1 \leq i < j \leq d}(\beta_i - \beta_j)\prod_{i=1}^{d}(\beta_i - 1) = 0$$

This cannot be true since β_i is irrational and $\beta_i \neq \beta_j$ for $i \neq j$. Thus, the result follows. ∎

Theorem 2.8 *The least period of the homogeneous recurrent sequence* (a) *of order k over the finite field F given by*

$$a_{n+k} = c_{k-1}a_{n+k-1} + \cdots + c_1 a_{n+1} + c_0 a_n + \alpha$$

divides the order of its companion matrix C in the group $\mathrm{GL}_k(F)$ if $c_0 \neq 0$.

Proof First of all, we need to check if C is invertible. Since $\det(C) = (-1)^{k-1}c_0 \neq 0$, C is invertible. Now, if $d = \mathrm{ord}(C)$ in $\mathrm{GL}_k(F)$, then $C^d = I_k$. If ρ is the least period of (a), then $a_{n+\rho} = a_n$ for all n. Using state vectors, we have $s_{n+\rho} = s_n$. Thus, $s_0 C^{n+\rho} = s_0 C^n$ and so $C^\rho = I_k$. Since ρ is the least period and d is a period of C, we have $\rho \mid d$. ∎

One question that arises is: when do we attain the maximum value of the least period of a homogeneous recurrent sequence? The answer to this question depends on the initial state vector. For a recurrent sequence (a) over a finite field F, we consider a duplicate (b) of (a) with the initial state vector $s_0 = (0, 0, \ldots, 1)$ which consists of $k - 1$ zeros. That is,

$$b_{n+k} = c_{k-1}b_{n+k-1} + \cdots + c_0 b_n$$

and $b_i = 0$ for $1 \leq i < k - 1$ and $b_{k-1} = 1$. So, the state vectors of (b) are

$$t_0 = \underbrace{(0, 0, \ldots, 0, 1)}_{k-1}$$

$$t_1 = \underbrace{(0, 0, \ldots, 0, 1, c_{k-1})}_{k-2}$$

$$\vdots$$

$$t_{k-1} = (1, c_{k-1}, \ldots)$$

This new sequence (b) is called the *impulse response sequence* of (a). The special property of this impulse sequence is that the first k state vectors form a basis of the vector space F^k. We can see that two state vectors t_a and t_b are same if and only if $C^a = C^b$. It follows from the fact that $t_n = t_0 C^n$.

Theorem 2.9 *Let (a) be a recurrent sequence over a finite field F and (b) be its impulse sequence. Then the least period of (a) divides the least period of (b).*

Proof Let τ and n_0 be the least period and pre-period of (b) respectively. By Theorem 2.5, it suffices to show that τ is a period of (a). Note that we have to consider the pre-period because we have not shown that (b) is periodic. So, for all $n \geq n_0$, we have $b_{n+\tau} = b_n$. Then $C^{n+\tau} = C^n$ for all $n \geq n_0$. This gives $s_{n+\tau} = s_n$ for all $n \geq n_0$, where s_i is the i-th state vector of (a). Thus, τ is a period of (a). ∎

In the above proof, if we assume that $c_0 \neq 0$, then (b) is periodic. In this case, the least period of (b) is the same as the order of C in $GL_k(F)$. In fact, the same is true if we replace the initial state vectors of the impulse sequence by any basis of F^k. It can be shown that (b) attains the maximum value of the least period as follows.

First, we see that the initial state vector of (a) can be written as a linear combination of the first k state vectors $t_0, t_1, \ldots, t_{k-1}$ of (b). Thus, every term of the sequence (a) is a linear combination of the first k state vectors of (b). Now, if λ is a period of recurrent sequences (x) and (y), then any linear combination of (x) and (y) also has period λ. Since τ is a period of (b), τ is a period of (a) as well. Thus, we get $\rho \mid \tau$.

We have to show now that $\tau = \text{ord}(C)$. Let d be the order of C in $GL_k(F)$. Since $C^d = I$, we get $t_d = t_0 C^d = t_0 I = t_0$. Thus d is a period of (b), whence $\tau \mid d$. Moreover, since τ is a period of (b), the state vectors t_i have period τ as well. Hence $t_{n+\tau} = t_n$, whence $t_n C^\tau = t_n$ for all n. This again implies $C^\tau = I$, so τ is a period of C. Thus, $d \mid \tau$.

The following theorem shows the connection between the order of characteristic polynomial and the period of a recurrent sequence.

Theorem 2.10 *Let F be a finite field and (a) be a recurrent sequence over F with companion matrix C and characteristic polynomial f. Then $\text{ord}(C) = \text{ord}(f)$ in $GL_k(F)$.*

Proof Since $f(x)$ is the monic characteristic polynomial of (a), it is the minimal polynomial of C over F. Let $d = \text{ord}(C)$ in $GL_k(F)$, so that $C^d = I_k$. Then $f(x)$ divides $x^d - 1$. Since d is the smallest such, by definition, $d = \text{ord}(f)$.   ∎

This yields the following theorem.

Theorem 2.11 *The impulse sequence of a recurrent sequence (a) over a finite field F has maximum least period $\text{ord}(C) = \text{ord}(f)$, where C is the companion matrix and f is the characteristic polynomial of (a) over F.*

We can use this theorem to say the following.

Theorem 2.12 *For any prime power m, there is a linear recurrent sequence (a) of order k over $\mathbb{F}_m$ with least period of $m^k - 1$.*

Proof By Theorem 1.6, $\mathbb{F}^*_{m^k}$ is cyclic of order $m^k - 1$. Let $f \in \mathbb{F}^*_{m^k}$ be a monic polynomial with $\mathbb{F}^*_{m^k} = \langle f \rangle$. By Theorem 2.11, if (a) is a recurrent sequence with characteristic polynomial f, then the impulse sequence (b) of (a) has least period $\text{ord}(f) = m^k - 1$.   ∎

We can combine these results into the following theorems.

Theorem 2.13 *Let F be a finite field and (a) be a homogeneous linear recurrent sequence of order k over F with characteristic polynomial $f(x) \in F[x]$. Then the least period of (a) is a divisor $\text{ord}(f)$ and the least period of the impulse sequence (b) of (a) is $\text{ord}(f)$. Both (a) and (b) are periodic if $f(0) \neq 0$.*

Theorem 2.14 *Let F be a finite field and (a) be a linear recurrent sequence of order k over F such that (a) has non-zero initial state vector and the characteristic polynomial $f(x) \in F[x]$ is irreducible. Then (a) is periodic and has least period $\text{ord}(f)$.*

Ward shows the necessity of this hypothesis using the following example. Consider the recurrence $a_{n+3} = a_{n+2} + 4a_{n+1} - 4a_n$ with characteristic polynomial $x^3 - x^2 - 4x + 4$, which factors as $(x - 1)(x + 2)(x - 2)$. For some specific initial values, we can write $a_n = 2^n + (-2)^n + c \cdot 1^n$ for some constant c. Note that $a_n = c$ for any odd n. Again, consider the recurrence $a_{n+4} = a_{n+2} + a_n$ with characteristic polynomial $x^4 - x^2 - 1$. This is irreducible but not cyclotomic (it has roots that are not roots of unity). But if $a_2 = a_4 = 0$ then $a_n = 0$ for all even n.

We have already proven that the linear recurrent sequence (a) given by (2.1) with non-zero c_0 is periodic and that (a) assumes the maximum least period when $a_i = 0$ for $0 \le i < k - 1$ and $a_{k-1} = 1$. [War33b] established a similar result. First he considers two linear recurrent sequences (a) and (b) modulo m that satisfies the same characteristic equation (so they also share the same characteristic polynomial). Then we can write (a) as a linear combination of (b) modulo m. For $0 \le i < k$, we write

$$x_1 a_i + \cdots + x_k a_{i+k-1} \equiv b_i \pmod{m}$$

Note that this is a system of linear congruences. It has an integer solution $x_1, x_2, \ldots,$ x_k modulo m if the determinant of the coefficient matrix is relatively prime to m. When $a_i = 0$ for $0 \le i < k - 1$ and $a_{k-1} = 1$, this determinant is equal to $(-1)^k$. So, such a solution exists and the least period of (a) is also a period of any (b) that satisfies the same recurrence. Earlier we said the same for any set of k initial state vectors that form a basis. We can obtain a similar result in this case.

Theorem 2.15 *Let (a) be a linear recurrent sequence of order k and*

$$
\Delta = \begin{vmatrix} a_0 & a_1 & \cdots & a_{k-1} \\ a_1 & a_2 & \cdots & a_k \\ \vdots & \vdots & \ddots & \vdots \\ a_{k-1} & a_k & \cdots & a_{2k-1} \end{vmatrix}
$$

If $\gcd(\Delta, m) = 1$ for a positive integer m, then the least period of (a) modulo m is the maximum least period among all such (a) that satisfy the same recurrence.

2.3 Fundamental Theorems on Periodicity

The next theorem shows that any linear recurrent sequence can be represented in terms of a null sequence and a periodic sequence.

Theorem 2.16 *Any linear recurrent sequence (a) can be uniquely written as a sum of a null and a periodic sequence modulo m. The periodic sequence has the same pre-period and least period as (a).*

[War33b, Page 601] explains the motivation behind such a representation using linear recurrences of order 1 as an example.

$$
a_{n+1} = \alpha a_n
$$

We can easily see that the solution of this equation is $a_n = a_0 \alpha^n$ for some integer a_0. So

$$
a_n = b_0 \alpha^n + c_0 \alpha^n
$$

where (b) is a null sequence and (c) is a periodic sequence, respectively, modulo n. Moreover, the pre-period and least period of (c) are the same as those of (a) since (b) does not affect either of them. The pre-period and least period of (a) can then be calculated as the least n such that

$$
b_0 \alpha^n \equiv 0 \quad (\text{mod } m) \tag{2.5}
$$

$$
c_0(\alpha^n - 1) \equiv 0 \quad (\text{mod } m) \tag{2.6}
$$

hold, respectively.

In particular, if $m = p$ for a prime p such that $p \nmid c_0$, then $n = \text{ord}_p(\alpha)$. So, even for order 1 we cannot find a general solution. For general m, we can calculate the pre-period and least period if we know the following:

(i) The prime factorizations of b_0, c_0, m and α. This is enough to calculate the pre-period.
(ii) $\text{ord}_p(\alpha)$, i.e., the least d such that

$$\alpha^d \equiv 1 \quad (\text{mod } p)$$

for a prime factor p of m. If m is square-free, this is enough to calculate the least period.
(iii) $\nu_p(\alpha^d - 1)$, i.e., the maximum exponent of p in $\alpha^d - 1$. This is unsolvable in general as it is equivalent to calculating $\nu_p(\alpha^{p-1} - 1)$.

Now, we can consider a general linear recurrent sequence with characteristic polynomial

$$f(x) = x^k - c_{k-1}x^{k-1} - \cdots - c_1 x - c_0$$

We will use *double modulus* for this purpose. The idea of the double modulus was introduced by Dedekind. Two integer polynomials $a(x)$ and $b(x)$ are congruent (mod $m, c(x)$) if and only if $a(x) - b(x)$ is divisible by $c(x)$ modulo m. That is,

$$a(x) - b(x) = c(x)q(x) + m \cdot r(x)$$

for some integer polynomials $q(x)$ and $r(x)$. The ordinary properties of congruences are satisfied by the double modulus when p is prime and $f(x)$ is irreducible. It acts like a prime modulus in the sense that if

$$a(x)b(x) \equiv 0 \quad (\text{mod } p, f(x))$$

then one of $a(x)$ or $b(x)$ is divisible by the modulus (mod $p, f(x)$). Thus, a congruence of double modulus can be divided by the same polynomial if it is not divisible by the modulus. Like residue classes, the polynomials which leave same remainder upon division by the double modulus are in the same class. The number of distinct residue classes (mod $p, f(x)$) is p^d where $d = \deg(f)$. Fermat's little theorem can be generalized for the double modulus as follows:

$$a(x)^{p^n} \equiv a(x) \quad (\text{mod } p, f(x))$$

for any integer polynomial $a(x)$. Similarly, the generalized form of Lagrange's theorem would be

$$\alpha^d + \alpha^{d-1}a(x) + \alpha^{d-2}b(x) + \cdots \equiv 0 \quad (\text{mod } p, f(x))$$

where the coefficients are integer polynomials has no more than d incongruent solutions (mod p, $f(x)$).

In a similar fashion as before, we can associate pre-period and least period with n such that

$$g(x)x^n \equiv 0 \quad (\text{mod } m, f(x))$$
$$h(x)(x^n - 1) \equiv 0 \quad (\text{mod } m, f(x))$$

Here, the polynomials $g(x)$ and $h(x)$ can be determined using the initial values $a_0, a_1, \ldots, a_{k-1}$. Then, as before, we can calculate the pre-period and least period if we know the following:

(i) The prime factorization of m.
(ii) The primary decomposition of $f(x)$, $g(x)$, and $h(x)$ modulo p^e, where $p^e \parallel m$ for a prime divisor p of m. The primary decomposition of a polynomial modulo p^e can be found using *Schönemann's second theorem*. This theorem states that if

$$f(x) \equiv c\phi_1(x)^{e_1} \cdots \phi_s(x)^{e_s} \quad (\text{mod } p)$$

where ϕ_i is an irreducible monic polynomial for $1 \leq i \leq s$, then there exists a decomposition of $f(x)$ modulo p^e

$$f(x) \equiv c' f_1(x) f_2(x) \cdots f_s(x) \quad (\text{mod } p^e)$$

where $f_i(x)$ is a monic polynomial such that

$$f_i(x) \equiv \phi_i(x)^{e_i} \quad (\text{mod } p)$$

(iii) For a prime divisor p of m and an irreducible polynomial factor $\phi(x)$ of $f(x)$ modulo p, the smallest positive integer d such that

$$x^d \equiv 1 \quad (\text{mod } p, \phi(x))$$

(iv) The polynomial $\ell(x)$ such that

$$x^d - 1 \equiv p\ell(x) \quad (\text{mod } p^2, \phi^2(x))$$

If $f(x)$ is the characteristic polynomial of the linear recurrent sequence (a) and ρ is the least period of (a), then Ward [War33a] says that (a) admits the period ρ (mod m, $f(x)$). Now, we can finally prove Theorem 2.16.

Proof Let n_0 and ρ be the pre-period and least period of (a), respectively, and let $n_0 \equiv -r \pmod{\rho}$ for some $0 \leq r < \rho$. Then $n_0 + r = k\rho$ for some integer k. Set $c_n = a_{\rho+r+n}$ and $b_n = u_n - c_n$ for $n \geq 0$. Then (c) is a periodic sequence with the same least period as (a) and

$$(a) = (b) + (c) \qquad\qquad (2.7)$$

On the other hand, (c) is a null sequence with pre-period n_0. Note that

$$
\begin{aligned}
c_{n+n_0} &= a_{n+n_0} - b_{n+n_0} \\
&= a_{n+n_0} - a_{n+k\rho+n_0} \equiv 0 \pmod{m} \\
c_{n_0-1} &= a_{n_0-1} - b_{n_0-1} \\
&= a_{n_0-1} - a_{k\rho+n_0-1} \not\equiv 0 \pmod{m}
\end{aligned}
$$

Now, if (2.7) is not unique then there is another representation $(a) = (v) + (w)$ with (w) a null sequence. Since $(c) - (w) = (v) - (b)$, we have that $(c) - (w)$ is a periodic null sequence. Thus, $(c) \equiv (w) \pmod{m}$ and so $(v) \equiv (b) \pmod{m}$.   ∎

Consider the polynomial

$$A_n(x) = a_0 x^{n-1} + a_1 x^{n-2} + \cdots + a_{n-1}$$

Then $f(x)A_n(x) = x^n P(x) - Q(x)$, where

$$
\begin{aligned}
P(x) &= a_0 x^{k-1} + (a_1 - c_{k-1}a_0)x^{k-2} + \cdots + (a_{k-1} - c_{k-1}a_{k-2} - \cdots - c_1 a_0) \\
Q(x) &= a_n x^{k+1} + (a_{n+1} - c_{k-1}a_n)x^{k-2} + \cdots + (a_{n+k-1} - c_{k-1}a_{n+k-2} - \cdots - c_1 a_n)
\end{aligned}
$$

Considering the identity modulo m,

$$x^n P(x) - Q(x) \equiv 0 \pmod{m, f(x)}$$

If (a) assumes the period n for a periodic (a), then

$$P(x) \equiv Q(x) \pmod{m}$$
$$(x^n - 1)P(x) \equiv \pmod{m, f(x)}$$

Conversely, if this congruence holds for some n, then (a) is periodic modulo m and has the period n. Now, assume that (a) is a null sequence modulo m with pre-period $n_0 \leq n$. Then

$$Q(x) \equiv 0 \pmod{m}$$
$$x^n Q(x) \equiv 0 \pmod{m, f(x)}$$

Similarly, if this holds for some n, then (a) is a null sequence modulo m with pre-period n. Thus, we obtain the following theorem of Ward.

Theorem 2.17 (Fundamental theorem on periodic sequences) *If (a) is a linear recurrent sequence of order k with characteristic polynomial $f(x)$, then a necessary and sufficient condition for (a) to be periodic with period ρ (mod m, $f(x)$) is*

$$(x^p - 1)P(x) \equiv 0 \quad (\mathrm{mod}\ m,\ f(x))$$

where $P(x)$ is a polynomial of degree $k - 1$ defined as

$$P(x) = a_0 x^{k-1} + (a_1 - c_{k-1}a_0)x^{k-2} + \cdots + (a_{k-1} - c_{k-1}a_{k-2} - \cdots - c_1 a_0)$$

The coefficients of $P(x)$ are completely determined by the initial values $a_0, a_1, \ldots,$ a_{k-1}.

Since this polynomial is of high interest in this theory, Ward calls it the *generator* of (a). Thus, we also have the following theorem of Ward.

Theorem 2.18 (Fundamental theorem on null sequences) *Let (a) be a linear recurrent sequence of order k with characteristic polynomial $f(x)$ and $P(x)$ be the generator of (a). Then a necessary and sufficient condition that (a) is a null sequence with pre-period $\leq n$ is*

$$x^n P(x) \equiv 0 \quad (\mathrm{mod}\ m,\ f(x))$$

2.4 Finding Pre-Period and Least Period

The generator of the recurrence (a) with initial values $0, \ldots, 0, 1$ is unity. So, we can also say the following about the least period of (a).

Theorem 2.19 *The least period of a linear recurrent sequence (a) modulo m with characteristic polynomial $f(x)$ is the least d such that*

$$x^d \equiv 1 \quad (\mathrm{mod}\ m,\ f(x))$$

We mentioned before that we can find the least period and pre-period given the factorization of some relevant integers and polynomials. Now we will focus on how we can find these values from the Schönemann decomposition. Note that, if $g = \gcd(a_0, a_1, \ldots, a_{k-1})$, then $g \mid a_n$ for $n \geq k$. So we can simply divide the whole sequence by g and get a reduced sequence. This allows us to establish the following theorem.

Theorem 2.20 *Let (a) be a linear recurrent sequence of order k and d be the greatest common divisor of the first k terms. Then the least period of (a) modulo m is same as the least period of (b) modulo $\frac{m}{\gcd(m,d)}$ where $b_n = \frac{a_n}{g}$.*

Let us again consider the Schönemann decomposition of $f(x)$. Let the decomposition of $f(x)$ modulo p be

$$f(x) \equiv f_1(x)^{r_i} \cdots f_s^{r_s}(x) \quad (\mathrm{mod}\ p)$$

Then there is a decomposition of $f(x)$ modulo p^e:

$$f(x) \equiv F_1(x) \cdots F_s(x) \quad (\text{mod } p^e) \tag{2.8}$$

where $F_i(x)$ are monic polynomials such that

$$F_i(x) \equiv f_i(x)^{r_i} \quad (\text{mod } p) \tag{2.9}$$

for $1 \leq i \leq s$. We can decompose the ring associated with the double modulus (mod p^e, $f(x)$) into the direct sum of the s rings corresponding to (mod p^e, $F_i(x)$). Take an element $u(x)$ from this ring so that

$$u(x) \equiv u_i(x) \quad (\text{mod } p^e, F_i(x))$$

where $\deg(u_i) < \deg(F_i)$. By the Chinese Remainder Theorem (CRT, see Chinese Remainder Theorem and Chinese Remainder Theorem for Polynomials in glossary), we can write $u(x)$ as a linear combination of $u_1(x), \ldots, u_s(x)$:

$$u(x) \equiv \beta_1(x)u_1(x) + \beta_2(x)u_2(x) + \cdots + \beta_s(x)u_s(x) \quad (\text{mod } p^e, f(x))$$

Here, we have $\deg(\beta_i) < \deg(f)$ and

$$\beta_i(x) \equiv 1 \quad (\text{mod } F_i(x))$$
$$\beta_i(x) \equiv 0 \quad (\text{mod } F_j(x)) \quad \text{if } i \neq j$$

If $u(x)$, $u_i(x)$ and $\beta_i(x)$ are generators of the linear recurrent sequences (u), (u_i) and (β_i), respectively, we have the analogous equation for (u):

$$(u) \equiv (\beta_1) \cdot (u_1) + (\beta_2) \cdot (u_2) + \cdots + (\beta_s) \cdot (u_s) \quad (\text{mod } p^e)$$

We have the following theorem.

Theorem 2.21 *Consider the Schönemann decomposition of $f(x)$ modulo p^e as discussed above where $f(x)$ is the characteristic polynomial of the linear recurrent sequence (a) of order k. Let $u(x)$ be a generator of (u) such that $\deg(u) < k$ and*

$$u(x) \equiv u_i(x) \quad (\text{mod } p^e, F_i(x))$$

where $\deg(u_i) < \deg(F_i)$ and $u_i(x)$ is a generator of (u_i) which is a linear recurrent sequence with characteristic polynomial $F_i(x)$. Then the least period of (u) (mod p^e, $f(x)$) is the least common multiple of the least periods of (u_i) (mod p^e, $F_i(x)$) and the pre-period is the maximum of the pre-periods of (u_i).

Returning to the periodicity of (a), assume that

$$(a) \equiv (b) + (c) \pmod{p^e}$$

is the decomposition of (a) into the null sequence (b) and the periodic sequence (a). Then for corresponding generator polynomials, we have

$$a(x) \equiv b(x) + c(x) \pmod{p^e, f(x)}$$

Moreover, considering the Schönemann decomposition of the generators,

$$a(x) \equiv a_i(x) \pmod{p^e, F_i(x)}$$
$$b(x) \equiv b_i(x) \pmod{p^e, F_i(x)}$$
$$c(x) \equiv c_i(x) \pmod{p^e, F_i(x)}$$

where $\deg(a_i), \deg(b_i), \deg(c_i) < \deg(F_i)$ and $a_i(x), b_i(x), c_i(x)$ are the generators of the recurrences $(a_i), (b_i), (c_i)$, respectively. So, we have

$$a_i(x) \equiv b_i(x) + c_i(x) \pmod{p^e, F_i(x)}$$

Now, if n_0 and ρ are the pre-period and least period of (a), respectively,

$$x^{n_0} b(x) \equiv 0 \pmod{p^e, f(x)}$$
$$x^\rho c(x) \equiv c(x) \pmod{p^e, f(x)}$$

This in turn gives us,

$$x^{n_0} b_i(x) \equiv 0 \pmod{p^e, F_i(x)}$$
$$x^\rho c_i(x) \equiv c_i(x) \pmod{p^e, F_i(x)}$$

Using what we proved, we can see that $b_i(x)$ is a null sequence and $c_i(x)$ is a periodic sequence. Thus, The pre-period of (b_i) and the least period of (c_i) is the pre-period and least period of (a_i), respectively. If ρ_i is the least period of (c_i) and n_i is the pre-period of (b_i), let $\rho = \text{lcm}(\rho_1, \rho_2, \ldots, \rho_s)$ and $n_0 = \max(n_1, n_2, \ldots, n_s)$. We show that ρ and n_0 are the least period and pre-period of (a), respectively.

$$(x^\rho - 1)c_i(x) \equiv 0 \pmod{p^e, F_i(x)}$$
$$(x^\rho - 1)c(x) \equiv 0 \pmod{p^e, f(x)}$$

This gives us that ρ is a period of (a) and by minimality of ρ, we get that ρ is the least period of (a). Similarly, we get the same result for n_0. Therefore, we have the following.

Theorem 2.22 *Let the prime factorization of m be $m = p_1^{e_1} \cdots p_r^{e_r}$ where $p_1, p_2, \ldots, p_r$ are distinct primes and (a) be a linear recurrent sequence. Then the pre-*

period of (a) *is the maximum of its pre-periods modulo* $p_i^{e_i}$ *and the least period of* (a) *is the least common multiple of all least periods of* (a) *modulo* $p_i^{e_i}$.

This theorem allows us to calculate the period of a linear recurrent sequence in terms of the smaller rings. Now, assume that c_0 in (2.1) is divisible by the prime p. Then we have the following theorem that gives us a criteria on pre-period.

Theorem 2.23 *Let* (a) *be a recurrent sequence of order* k *and* $f(x)$ *be its characteristic polynomial. If the decomposition of* $f(x)$ *is*

$$f(x) \equiv F_1(x) \cdots F_s(x) \quad (\mathrm{mod}\ p^e)$$

and c_0 *in (2.1) is divisible by* p *so that* $F_1(x) \equiv x^t + pq(x)$ *for some integer polynomial* $q(x)$, *then a necessary and sufficient condition that* (a) *is a null sequence modulo* p^e *is that*

$$u(x) \equiv 0 \quad (\mathrm{mod}\ p^e, F_2(x) \cdots F_s(x))$$

where $u(x)$ *is the generator polynomial of* (a).

Proof Consider the decomposition of $f(x)$. Since $p \mid c_0$, there is an i such that $F_i(x) = x^t + pq(x)$ for some integer polynomial $q(x)$. Without loss of generality, it was assumed to be $F_1(x)$ in the statement. This exponent t is actually the number of consecutive coefficients in (2.1) which is divisible by p. For brevity, let

$$g(x) = F_2(x) \cdots F_s(x)$$

Here, we should have that $\Re(F_1(x), g(x))$ is not divisible by p. By the fundamental theorems on null and periodic sequences, (u) is a null sequence modulo p^n if and only if

$$x^n u(x) \equiv 0 \quad (\mathrm{mod}\ p^e, f(x))$$

has a solution. This congruence has a solution if and only if the following two congruences have solutions.

$$x^n u(x) \equiv 0 \quad (\mathrm{mod}\ p^e, F_1(x))$$
$$x^n u(x) \equiv 0 \quad (\mathrm{mod}\ p^e, g(x))$$

The first congruence has solution for any $u(x)$ if we consider $n = mt$. But the second one is solvable if and only if $u(x) \equiv 0 \pmod{p^e g(x)}$ when $\gcd(p, \Re(x, g(x))) = 1$. Thus, we have the theorem.   ∎

We can see that we can find the pre-period by finding the smallest n such that

$$x^n u(x) \equiv 0 \quad (\mathrm{mod}\ p^e, F_1(x))$$

where $F_1(x)$ is the polynomial we considered in the proof. In a similar fashion, we can also prove the analogous result for period of (a).

Theorem 2.24 *With same notation and terminology in Theorem 2.23, a necessary and sufficient condition that (a) is periodic modulo p^e is*

$$u(x) \equiv 0 \pmod{p^e, F_1(x)}$$

Given these two theorems, we can now decompose (a) into a null sequence and a periodic sequence as claimed. Since $p \nmid \Re(F_1(x), g(x))$, we can find two polynomials $\alpha(x), \beta(x)$ such that

$$\alpha(x) F_1(x) + \beta(x) g(x) \equiv u(x) \pmod{p^e, f(x)}$$

Then we see that letting $\beta(x) g(x) = b(x)$ and $\alpha(x) F_1(x) = c(x)$, the degrees of both α and β are less than k. If (b) and (c) are the recurrent sequences generated by $b(x)$ and $c(x)$, respectively,

$$u(x) \equiv b(x) + c(x) \pmod{p^e, f(x)}$$
$$(a) \equiv (b) + (c) \pmod{p^e}$$

Here, (b) is a null sequence modulo p^e and (c) is a periodic sequence modulo p^e. If $u(x)$ is the generator of (a), then $u(x)$ can be written as

$$u(x) \equiv v(x) F_2(x) \cdots F_s(x) \pmod{p^e}$$

and if the pre-period is n_0,

$$x^n v(x) \equiv 0 \pmod{p^e, F_1(x)}$$

where $F_1(x) = x^t + p^r \cdot q(x)$ for an integer polynomial $q(x)$ and a positive integer r where $p \nmid q(x)$ and $\deg(q) < t$. Using Schönemann's decomposition on $u(x)$,

$$v(x) \equiv p^\alpha v_1(x) w(x) \pmod{p^e}$$

such that $v_1(x)$ is of the form $x^{t_1} + p^r q_1(x)$ where $p \nmid q_1(x)$ and $\deg(q_1) < t_1$. Again, it follows that the pre-period n_0 is the smallest n such that

$$x^n v_1(x) \equiv 0 \pmod{p^{e-\alpha}, F_1(x)}$$

This leads to a recursive process of finding pre-period and by inducting, we get the following theorem.

Theorem 2.25 *Let the polynomials $u(x), v(x), w(x)$ be defined as*

$$u_0(x) = v_1(x)u'(x)$$
$$u_0'(x) = u'(x)$$
$$u_{n-1}(x) \equiv v_n(x)u_{n-1}'(x) \quad (\mathrm{mod}\ p^{e_{r-1}})$$
$$x^{t-t_n}v_n(x) \equiv p^{\alpha_n}u_n(x) \quad (\mathrm{mod}\ F_1(x))$$
$$v_n(x) = x^{e_r} + p^{\beta_n}w_n(x)$$
$$e_n = e - \alpha - \alpha_1 - \cdots - \alpha_n$$

Here, $u_n(x)$ and $u_{n-1}'(x)$ are not divisible by x modulo p, $\deg(w_n) < t_n$. Moreover, α_i is positive and so we will find a positive integer m such that we have either

$$e \le \alpha + \alpha_1 + \cdots + \alpha_m \quad \text{or}$$
$$\mathfrak{R}(u_m(x), F_1(x)) \not\equiv 0 \quad (\mathrm{mod}\ p)$$

Consider the smallest of such m. Then, the pre-period n_0 of (a) is,

$$n_0 = \begin{cases} mt - (t_1 + \cdots + t_m) \text{ in the first case} \\ mt + l - (t_1 + \cdots + t_m) \text{ in the second case} \end{cases}$$

where l is the smallest value of n for which

$$x^n \equiv 0 \quad (\mathrm{mod}\ p^{e_m}, F_1(x))$$

We can also prove the following by induction.

Theorem 2.26 *Let $u(x)$, $v(x)$ and $w(x)$ be polynomials defined as*

$$u_n(x) \equiv (x^{\alpha_n} + p^{\beta_n}w(x))v(x) \quad (\mathrm{mod}\ p^{e_n})$$
$$x^{t-\alpha_n}u_n(x) \equiv p^{\gamma_n}u_{n+1}(x) \quad (\mathrm{mod}\ F_1(x))$$

Here, $p \nmid u_n(x)$, $v(x)$ is not divisible by x modulo p, $\deg(w_n(x)) < \alpha_n$, where α_n is the number of consecutive coefficients of x^i in $u_n(x)$ which are divisible by p. Since γ_i is positive, we will again find h such that

$$e_m \le \gamma_1 + \cdots + \gamma_h \text{ or}$$
$$\alpha_h = 0$$
$$\mathfrak{R}(u_h(x), F_1(x)) \not\equiv 0 \quad (\mathrm{mod}\ p) \text{ otherwise}$$

Consider the smallest such h. Then, in the first case, the least value of n for which we have $x^n \equiv 0 \ (\mathrm{mod}\ p^{e_m}, F_1(x))$ is

$$n_0 = ht - (\alpha_1 + \alpha_2 + \cdots + \alpha_h)$$

In the second case,

$$n_0' = \ell(h)n_0$$

$$\ell(h) = \left\lceil \frac{e_m}{\alpha_1 + \cdots + \alpha_h} \right\rceil$$

where n_0 is exactly the quantity calculated in the first case and $\lceil x \rceil$ is the ceiling of x.

The following results can be derived using these theorems.

Theorem 2.27 *Let (a) be a linear recurrent sequence and let the Schönemann factorization of its characteristic polynomial $f(x)$ be*

$$f(x) \equiv F_1(x) \cdots F_s(x) \pmod{p^e}, \quad \text{where}$$
$$F_1(x) \equiv x^t - p^\alpha u(x), \quad p \nmid u(x)$$

Then the least upper bound of the pre-period of any such (a) modulo p^e is $t \left\lceil \frac{e}{\alpha} \right\rceil$.

Theorem 2.28 *Let (a) be a linear recurrent sequence such that the last t coefficients are divisible by p. Then the least upper bound of the pre-period of (a) modulo p^e is et.*

Theorem 2.29 *The least upper bound of the pre-period of any linear recurrent sequence of order k modulo p^e is ek.*

We want to develop similar theorems for period of (a). In order to do that, we will have to discuss some facts about polynomial congruence of double modulus. [War33c] showed some fundamental results on this. Let us consider the following problem. Assume that

$$f(x) = x^k + c_{k-1}x^{k-1} + \cdots + c_1 x + c_0$$

and m is an integer relatively prime to c_0. If $A(x)$ is a polynomial such that

$$A(x) \not\equiv 0 \pmod{m, f(x)}$$

then we want to find all polynomials $B(x)$ such that

$$A(x)B(x) \equiv 0 \pmod{m, f(x)} \tag{2.10}$$

One can think of this as a generalization of the classical problem of finding x satisfying $ax \equiv 0 \pmod{m}$, given $a \not\equiv 0 \pmod{m}$. First, considering the prime factorization of m, we can reduce this problem to the case when m is a prime power. We can use Schönemann's second theorem and the Chinese Remainder Theorem for this.

$$f(x) \equiv c_0 f_1(x)^{e_1} \cdots f_s(x)^{e_s} \pmod{p}$$
$$f(x) \equiv c_0' F_1(x) \cdots F_s(x) \pmod{p^e}$$
$$F_i(x) \equiv f_i(x)^{e_i} \pmod{p}$$

Here, $\mathfrak{R}(F_i, F_j)$ is not divisible by p. Let the prime factorization of m be $m = p_1^{e_1} \cdots p_r^{e_r}$. Consider the congruences

$$A(x)B(x) \equiv 0 \pmod{p_i^{e_i}, F_i(x)}$$

Then the Chinese Remainder Theorem ensures that there is a solution to the congruence

$$A(x)B(x) \equiv 0 \pmod{p_1^{e_1} \cdots p_r^{e_r}, F_1(x) \cdots F_s(x)}$$

Now, with regards to the solution of congruence (2.10), we can assume without loss of generality that $\deg(B) < \deg(f)$. For such an arbitrary solution $B(x)$, let $p^\alpha \parallel B(x)$. If $e \le \alpha$, there is a solution $B(x) \not\equiv 0 \pmod{p}$. Otherwise, we can reduce congruence (2.10) to a congruence of the same form. Consider the resultant $\mathfrak{R}(A, f)$ and the Sylvester matrix $C(A, f)$ of A and f

$$C = \begin{pmatrix} a_d & a_{d-1} & \cdots & a_0 \\ 1 & c_{k-1} & \cdots & c_0 \end{pmatrix}$$
$$A(x) = a_d x^d + a_{d-1} x^{d-1} + \cdots + a_0$$

Let $C = (c_{ij})$ be the transpose of C. Assume that $\det(C) = p^l C'$ where $\gcd(p, C') = 1$ and $p^l \parallel \det(C)$. Let us rewrite congruence (2.10) as

$$A(x)B(x) + f(x)V(x) = p^e W(x)$$

such that $\deg(V) < d$ and $\deg(W) < d + k$. We can compare coefficients of both side and get a system $d + k$ linear equations.

$$\sum_{i=1}^{d+k} r_i x_i = p^e w_i$$

where $x_1, x_2, \ldots, x_{d+k}$ are the unknown coefficients of $B(x)$ and $V(x)$ and $w_1, w_2, \ldots, w_{d+k}$ are the coefficients of $W(x)$. We have that the determinant of this system is $\det(C)$. Then

$$\det(C)x_j = p^e \sum_{i=1}^{d+k} \bar{r}_{ji} w_i$$

where $\bar{r}_{ji}$ is the co-factor of r_{ij} in C. Assume that $h = \min \nu_p(\bar{r}_{ji})$, so h is highest power of p that divides the co-factors. Since $\det(C) = p^l C'$, letting $\bar{r}_{ji} = p^h r'_{ji}$,

$$C' x_j = p^{e+h-l} \sum_{i=1}^{m+n} r'_{ji} w_i$$

Finally, there must be a t such that $p \nmid r'_{jt}$. Then setting $w_t = 1$ and $w_j = 0$ for other j, we have a solution of the linear equation system such that x_i is divisible by p^{e+h-l}. There is at least one x_i such that $p^{e+h-l} \| x_i$. Let us call $p^{l-h} = p^{h'}$ the *first elementary divisor* of C.

Theorem 2.30 *The least value of e such that any solution $B(x)$ of (2.10) such that $\deg(B) < \deg(f)$ is divisible by $p^{h'}$ where $p^{h'}$ is the first elementary divisor with respect to the prime p of $C(A, f)$. If $e > h'$, setting $B(x) = p^{e-h'} W(x)$,*

$$A(x) W(x) \equiv 0 \pmod{p^{h'}, f(x)}$$

We can use this to prove the next theorem.

Theorem 2.31 *Let (a) be a linear recurrent sequence of order k generated by $a(x)$ with characteristic polynomial $f(x)$. If the first elementary divisor of the Sylvester matrix of $C(a, f)$ is p^l, then for $e > l$, the least period of (a) (mod p^e, $f(x)$) is a multiple of the least period of (a) modulo p^{e-l}.*

Proof Let ρ be the least period of (a) (mod p^e, $f(x)$).

$$(x^\rho - 1)a(x) \equiv 0 \pmod{p^e, f(x)}$$

Since p^l is the first elementary divisor of $C(a, f)$,

$$x^\rho - 1 \equiv 0 \pmod{p^{e-l}, f(x)}$$

Thus, ρ is a multiple of the least period of (a) (mod p^{e-l}).   ∎

This theorem is very important for the calculation of least period. From the results established before, we can consider only (a) (mod p^e) for determining the least period of (a).

Theorem 2.32 *Let $a(x)$ be the generator of a linear recurrent sequence (a) of order k with characteristic polynomial $f(x)$ such that*

$$a(x) \equiv \alpha(x)^u \pmod{p}$$
$$f(x) \equiv \alpha(x)^v g(x) \pmod{p}$$

where $\alpha(x)$ is a monic irreducible polynomial modulo p and $\mathfrak{R}(g, \alpha)$ is not divisible by p. Then the least period of (a) modulo p is $p^e l$ where e is the unique integer such that $p^{e-1} < u - v \le p^e$ and l is the least value of n such that

$$x^n - 1 \equiv 0 \quad (\mathrm{mod} \; p, \, f(x))$$

Proof Recall that the characteristic polynomial $f(x)$ can be written in the form

$$f(x) = \alpha(x)^u - p\beta(x)$$

where $\alpha(x)$ is a monic irreducible polynomial modulo p and $\deg(\beta) < \deg(f) = k$. Using the fundamental theorems, we can assume that (a) is periodic without loss of generality. Then, there is a n such that

$$(x^n - 1)a(x) \equiv 0 \quad (\mathrm{mod} \; p^e, \, f(x))$$

where $a(x)$ is the generator of (a). By the theorem above, we have already established that if ρ is the least period and

$$(x^\rho - 1)u(x) \equiv 0 \quad (\mathrm{mod} \; p^e, \, f(x))$$

and p^l is the first elementary divisor, then

$$x^\rho - 1 \equiv 0 \quad (\mathrm{mod} \; p^{e-l}, \, f(x))$$

Moreover, ρ is a multiple of the least period of (a) modulo $p^{\rho-l}$. Since $u(x)$ is not trivially divisible by p, we can assume without loss of generality that

$$u(x) \equiv \alpha(x)^v g(x) \quad (\mathrm{mod} \; p)$$

where $u > v$ and $p \nmid \mathfrak{R}(\alpha, g)$. Again, using Schönemann's second theorem,

$$u(x) \equiv u'(x)g(x) \quad (\mathrm{mod} \; p^e)$$

so that

$$u'(x) \equiv \alpha(x)^v + p\beta(x)$$

where $\deg(\beta) < \deg(u')$ and $g(x) \equiv \beta(x) \pmod{p}$. Then the least period of (a) is the least positive integer n such that

$$(x^n - 1)u'(x) \equiv 0 \quad (\mathrm{mod} \; p^e, \, f(x)) \tag{2.11}$$

If $e = 1$, then (2.11) can be replaced with

$$x^n - 1 \equiv 0 \quad (\mathrm{mod} \ p, \alpha(x)^{u-v}) \tag{2.12}$$

Let d be the degree of $\alpha(x)$. Then the least period of

$$x^n - 1 \equiv 0 \quad (\mathrm{mod} \ p, \alpha(x))$$

is a divisor of $p^d - 1$ by Theorem 2.12. Let us denote this divisor by l. Note that we can also have $l = p^d - 1$ for some recurrent sequences. Then

$$x^l - 1 = \alpha(x)g(x) + p\beta(x)$$
$$x^l - 1 \equiv \alpha(x)g(x) \quad (\mathrm{mod} \ p)$$

such that $\deg(\beta) < \deg(\alpha)$. The discriminant of $x^l - 1$ is relatively prime to p, so

$$g(x) \not\equiv 0 \quad (\mathrm{mod} \ p, \alpha(x))$$

Then we have

$$x^{rl} - 1 \equiv r\alpha(x)g(x) \quad (\mathrm{mod} \ p, \alpha(x)^2)$$

Thus, the least period of

$$x^n - 1 \equiv 0 \quad (\mathrm{mod} \ p, \alpha(x)^2)$$

if pl. Furthermore, we have

$$x^{pl} - 1 \equiv \alpha(x)^p g(x)^p \quad (\mathrm{mod} \ p)$$

so pl is the least period of (2.12) if $2 \le u - v \le p$. Continuing in this fashion (or we can use induction but that might be too tedious), we obtain the theorem. ■

Thus, we also have the following theorem with the same notations and definitions.

Theorem 2.33 *The least period of (a) modulo p is $p^r l$ where r is the unique integer such that $p^{r-1} < u < p^r$ and the least upper bound of the least period is $p^r(p^d - 1)$.*

We only checked the case modulo p in the theorem above. Now, we will check the case $m = p^e$ for a positive integer $e > 1$. We can assume that

$$x^{p^r l} - 1 \equiv p^\gamma g(x) \quad (\mathrm{mod} \ f(x)) \tag{2.13}$$

such that $p \nmid g(x)$. Then using $g(x) \not\equiv 0 \ (\mathrm{mod} \ p)$,

$$u'(x)g(x) \equiv 0 \quad (\mathrm{mod} \ p)^\tau h(x) \quad (\mathrm{mod} \ f(x)) \tag{2.14}$$

where τ is the largest positive integer such that $h(x) \not\equiv 0 \pmod{p}$ and τ is the largest positive integer n such that

$$u(x)g(x) \equiv 0 \pmod{p^n, f(x)}$$

From (2.13),

$$x^{p^r l} \equiv (1 + p^\gamma g(x) \pmod{f(x)}$$
$$x^{p^{r+t} l} \equiv (1 + p^\gamma g(x))^{p^t} \pmod{f(x)}$$
$$\equiv 1 + \binom{p^t}{1} p^\gamma g(x) + \binom{p^t}{2} p^{2\gamma} g(x)^2 + \cdots + p^{\gamma p^t} g(x)^{p^t} \pmod{f(x)}$$
$$x^{p^{r+t} l} - 1 \equiv (1 + p^\gamma g(x))^{p^t} \pmod{f(x)}$$
$$\equiv p^{\gamma+t} g(x) + \frac{p^t - 1}{2} p^{2\gamma+t} g(x)^2 + \cdots + p^{\gamma p^t} g(x)^{p^t} \pmod{f(x)}$$

Multiplying both side with $u'(x)$ and using (2.14),

$$u'(x)(x^{p^{r+t} l} - 1) \equiv u'(x) p^{\gamma+t} g(x) + u'(x) \frac{p^t - 1}{2} p^{2\gamma+t} g(x)^2 + \cdots + u'(x) p^{\gamma p^t} g(x)^{p^t} \pmod{f(x)}$$
$$\equiv p^{\gamma+t+\tau} h(x) + \frac{p^t - 1}{2} p^{2\gamma+t+\tau} g(x)h(x) + \cdots + p^{\gamma p^t + \tau} g(x)^{p^t - 1} h(x) \pmod{f(x)}$$

Thus, we get the double modulus congruence

$$u'(x)(x^{p^{r+t} l} - 1) \equiv p^{\gamma+\tau+t} h(x) \pmod{p^{\gamma+\tau+t+1}, f(x)}$$

The congruence may not hold for a trivial case $p = 2$ and $\gamma = 1$ so we will exclude it for now. Using the notations above, we have the following theorems.

Theorem 2.34 *Let p be an odd prime and $e > 1$ be a positive integer. Then the least period of (2.11) is $p^r l$ if $e \le \gamma + \tau$ and $p^{r+e-\gamma-\tau} l$ if $e \ge \gamma + \tau$.*

Theorem 2.35 *Let p be an odd prime. Then the least upper bound of least period for any $u'(x)$ is $p^r l$ if $e \le \gamma$ and $p^{r+e-\gamma}$ if $e \ge \gamma$.*

Inspired by Theorem 2.12, we also get the following result.

Theorem 2.36 *Let (a) be a linear recurrent sequence of order k and ℓ be the least common multiple of $1, 2, \ldots, k$. If $\mathcal{D}$ is the discriminant of the characteristic polynomial $f(x)$ and $p \nmid \mathcal{D}$, then $p^\ell - 1$ is a period of (a) modulo p.*

The following theorem is also trivial at this point. And we can use it to prove the theorem above.

Theorem 2.37 *Let p be a prime and the factorization of $f(x)$ modulo p in irreducible polynomials be*

$$f(x) \equiv f_1(x)^{e_1} \cdots f_s^{e_s}(x) \quad (\mathrm{mod}\ p)$$

Then the least period ρ of (a) which $f(x)$ is the characteristic polynomial of, is the least common multiple of $p^{e_i} - 1$.

We will use these theorems to establish some more results regarding divisibility sequences and linear recurrent sequences. There is a lot we can say about recurrent sequences, especially on Ward's account. But for now, we end our discussion of recurrent sequence here.

References

[Bel30]   E.T. Bell, Periodic recurring series. Proc. Natl. Acad. Sci. **16**(11), 750–752 (1930). https://doi.org/10.1073/pnas.16.11.750

[Car29]   R.D. Carmichael, A simple principle of unification in the elementary theory of numbers. Am. Math. Mon. **36**(3), 132–143 (1929). ISSN: 00029890, 19300972. http://www.jstor.org/stable/2299678

[War33a] M. Ward, A property of recurring series. Proc. Natl. Acad. Sci. USA **19**(10), 914–916 (1933). ISSN: 0027-8424. https://doi.org/10.1073/pnas.19.10.914. eprint: https://www.pnas.org/content/19/10/914.full.pdf, https://www.pnas.org/content/19/10/914

[War33b] M. Ward, The arithmetical theory of linear recurring series. Trans. Am. Math. Soc. **35**(3), 600–628 (1933). ISSN: 00029947, http://www.jstor.org/stable/1989851

[War33c] M. Ward, The cancellation law in the theory of congruences to a double modulus. Trans. Am. Math. Soc. **35**(1), 254–260 (1933). ISSN: 00029947, http://www.jstor.org/stable/1989323

Chapter 3
Divisibility Sequences

3.1   Introduction

Consider the classic problem that the product of n consecutive integers is divisible by $n!$. The proof of this fact is the basis of our study on this topic. A beginner usually tries to prove this with some basic modular arithmetic, for example, at least one of the n consecutive integers is divisible by n since each of them leaves a different remainder upon division by n. Similarly, at least one of those integers is divisible by i for $1 \leq i \leq n$. However, this does not prove that the product of all $1 \leq i \leq n$ divides $n!$ as well, although the least common multiple of them $\operatorname{lcm}(1, 2, \ldots, n)$ does. The most natural way to prove this claim is to write it like this:

$$\frac{k(k-1)\cdots(k-n+1)}{1 \cdot 2 \cdots n} = \frac{1 \cdot 2 \cdots k}{1 \cdot 2 \cdots n \cdot 1 \cdot 2 \cdots (k-n)}$$

$$= \frac{k!}{n!(k-n)!}$$

$$= \binom{k}{n}$$

This is the binomial coefficient $\binom{k}{n}$, the number of ways to choose n distinct objects from k distinct objects, which is certainly an integer. This problem alone does not tell us much. But consider the following problem next.

Problem 3.1 Prove that the product of k consecutive Fibonacci numbers is divisible by the product of the first k consecutive Fibonacci numbers.

In 2014, the first author posed the following problem to Nur Muhammad Shafiul-lah Mahi at the national math camp of Bangladesh (he was an IMO contestant for Bangladesh from 2011 to 2014 and earned bronze medals at 2012, 2013 and a silver medal at 2014, respectively).

© The Author(s), under exclusive license to Springer Nature Singapore Pte Ltd. 2021
M. Billal and S. Riasat, *Integer Sequences*,
https://doi.org/10.1007/978-981-16-0570-3_3

Problem 3.2 Show that for any positive integers n and x,

$$(x - 1)(x^2 - 1) \cdots (x^{2014} - 1) \mid (x^n - 1)(x^{n+1} - 1) \cdots (x^{n+2013} - 1)$$

holds.

The common denominator in all three problems is that the product of k consecutive terms in their sequence is divisible by the product of the first k terms. So, what other sequences are there which maintain the same property? A bigger question is, when can a sequence have this property? In order to understand the underlying structure of such sequences, we will try to find the common properties of all three sequences. First, recall two important properties that both (F) and $(x^n - 1)$ have. If $m \mid n$, then $F_m \mid F_n$ and $x^m - 1 \mid x^n - 1$. In fact, this can be generalized to $\gcd(F_m, F_n) = F_{\gcd(m,n)}$ and $\gcd(x^m - 1, x^n - 1) = x^{\gcd(m,n)} - 1$. These observations lead us to an answer to the question above which we will shortly answer.

Following Hall [Hal36], we will call an integer sequence (a) a *divisibility sequence* if for any positive integer m, n we have $a_m \mid a_n$ whenever $m \mid n$. Morgan Ward also called this *Lucasian sequence* but that term was dropped afterwards. In order to make our lives simpler, we will consider only sequences of positive integers in this text so that we do not have to deal with absolute values every time there is such a possibility.

This property can be strengthened changing the condition to $\gcd(a_m, a_n) = a_{\gcd(m,n)}$. The sequence of the latter kind is called *strong divisibility sequence* [Now15]. The reason is that the second property is a generalization of the first one. Any strong divisibility sequence is also a divisibility sequence due to the fact $\gcd(m, n) = m$ if $m \mid n$ which implies $\gcd(a_m, a_n) = a_{\gcd(m,n)} = a_m$ or $a_m \mid a_n$. Note that a divisibility sequence is not necessarily a strong one. For example, Carmichael's *universal exponent function* $\lambda(n)$ defined as

$$\lambda(n) = \begin{cases} 1 & \text{if } n = 2 \\ 2^{k-2} & \text{if } n = 2^k \\ p^{k-1}(p - 1) & \text{if } n = p^k \\ \text{lcm}(\lambda(p^e), \lambda(m)) & \text{if } n = p^e m \text{ with } p \nmid m \end{cases}$$

is a divisibility sequence but not a strong one. We can easily verify that $\lambda(a) \mid \lambda(b)$ if $a \mid b$. We can even see that $\text{lcm}(\lambda(a), \lambda(b)) = \lambda(\text{lcm}(a, b))$. But $\gcd(\lambda(a), \lambda(b))$ is not necessarily $\lambda(\gcd(a, b))$. We will discuss this function more in Sect. 5.2. In a similar fashion, $a_n = \varphi(n)$ where $\varphi(n)$ is Euler's totient function also forms a divisibility sequence but not a strong one. $a_n = n!$ also forms a divisibility sequence but not a strong one since $m \mid n$ implies $m \le n$ and $m! \le n!$ so $m! \mid n!$. But $\gcd(m!, n!) = m!$ which is not necessarily $\gcd(m, n)!$.

For a strong divisibility sequence (a), Ward [War55] called (c) a *subsequence* of (a) for a fixed positive integer s if

$$c_n = \frac{a_{sn}}{a_s}$$

for all $n \in \mathbb{N}$. We set $a_0 = 0$ for any divisibility sequence (a). Also, note that we can assume without loss of generality that $a_1 = 1$. Because $a_1 \mid a_n$ for any n, we can divide the whole sequence by a_1 and still get a divisibility sequence. In this regard, we call n a *trivial divisor* of (a) if $n \mid a_i$ for all $i \in \mathbb{N}$ because of the same reason. Therefore, if we ever encounter a situation where we have a trivial divisor n, we can assume without loss of generality that $n = 1$. Moreover, in order to keep a divisibility sequence normalized, we can assume without loss of generality that $a_0 = 0$ and $a_1 = 1$. Marshal Hall [Hal36] called such divisibility sequences *normal*. However, as we can assume so without loss of generality, we will freely do so without any restriction of assigning a name for such a case.

For a sequence (a), we define a sequence (c) as

$$c_n = \frac{\operatorname{lcm}(a_1, a_2, \dots, a_n)}{\operatorname{lcm}(a_1, a_2, \dots, a_{n-1})}$$

Following Nowicki [Now15], (c) will be called the *lcm sequence* of (a). If (c) is the lcm sequence of (a), then (c) is a divisibility sequence but not a strong one.

We call (a) a *binomial sequence* if for any non-negative integer k and n,

$$\binom{n}{k}_a = \frac{a_n a_{n-1} \cdots a_{n-k+1}}{a_1 a_2 \cdots a_k}$$

is an integer. We can also call $\binom{n}{k}_a$ the binomial coefficient of (a). In other words, (a) is a binomial sequence if the binomial coefficient of (a) is always an integer. For $a_n = n$, we get $\binom{n}{k}_a = \binom{n}{k}$, the usual binomial coefficient. Ward used the notation $[n, k]$ to denote the binomial coefficient. Some authors used $\left[\begin{smallmatrix} n \\ k \end{smallmatrix} \right]$. We have opted for $\binom{n}{k}_a$ because it is consistent with the usual binomial coefficient notation and it bears the context of (a). Moreover, we denote the product of first n terms of (a) using $n!_a$, following the footprints of binomial coefficients and factorials. For $k > n$, we have $\binom{n}{k}_a = 0$.

Let us call a sequence (a) an *odd divisibility sequence* if $a_m \mid a_{mn}$ for all odd n. So, a divisibility sequence is an oddly divisibility sequence but the converse does not necessarily have to be true. A good demonstration of this would be to consider $a_n = 2^n - 1$. In fact we could simply consider $a_n = x^n - y^n$ due to the fact $\gcd(x^m - y^m, x^n - y^n) = x^{\gcd(m,n)} - y^{\gcd(m,n)}$ for any positive integers m, n. On the other hand, if $a_n = 2^n + 1$, then (a) is not a divisibility sequence even though it is an oddly divisibility sequence. It follows from the fact that $a + 1$ divides $a^k + 1$ for any odd k. We could impose the condition that $a_m \mid a_{mn}$ holds only for odd n and no other n. But such imposition seems unnecessary and hence, we can allow divisibility sequences to be oddly divisibility sequences.

For a positive integer b, if the highest power of a that divides b is r, we write $a^r \| b$ or $\nu_a(b) = r$. For a particular sequence (a) and a positive integer m, m divides (a) if there exists an index r such that $m \mid a_r$. The smallest such index r is called the *rank of apparition* of m in (a). Let $\rho(p, e)$ be the smallest index for which $a_{\rho(p,e)}$

is divisible by p^e. For $e = 1$, we will simply write $\rho(p)$ instead of $\rho(p, 1)$. If the context is clear, we will use ρ instead of $\rho(p)$.

Note that, we do not call ρ the period of (a) modulo p yet. Because the sequence does not necessarily have to repeat after a_ρ if $p \mid a_\rho$. The condition for the sequence to repeat modulo p is that there must be a k such that $a_k \equiv a_0 \pmod{p}$ and $a_{k+1} \equiv a_1 \pmod{p}$. Only then we have the sequence (a) periodic modulo p since then $a_{k+2} = aa_{k+1} - ba_k \equiv av - bu \equiv a_2 \pmod{p}$ and the sequence is repeated.

Morgan Ward discussed divisibility sequences in a lot of his papers. As we will see later, divisibility sequences are intimately related to recurrent sequences. Ward derived a lot of beautiful results which can be found today in the literature, only in a modern form so to say. For example, the connection between a strong divisibility sequence and lcm sequence was discovered by Ward. He stated that if $a_n = \prod_{d\mid n} b_d$, then

$$b_n = \frac{a_n}{\text{lcm}\left(a_{\frac{n}{p_1}}, a_{\frac{n}{p_2}}, \ldots, a_{\frac{n}{p_k}}\right)}$$

where $p_1, p_2, \ldots, p_k$ are distinct prime divisors of n. An equivalent result was published in 2015 [Now15]. This goes to show how timeless Ward's results are. Lucas [Luc78a, Luc78b, Luc78c] discussed his sequences (U) and (V) in details. Then Lehmer [Leh30] generalized Lucas's work. Lucas only used some special cases of what we could generalize even further. We will see this as a general Lucas sequence in Chap. 4. We will see if Lucas missed whether there were more sequences like (U) and (V) or were those the only ones. For example, (U) is a strong divisibility sequence. We will check if there are other sequences possible which are divisibility sequences as well. Binomial coefficients can be generalized based on divisibility sequences. We will explore the connection between them.

We start by characterizing divisibility sequences. To be more precise, we will characterize strong divisibility sequences by their divisors. But first, let us check some well-known divisibility sequence. Obviously, (a) with $a_n = n$ is a strong divisibility sequence with binomial coefficients being the traditional $\binom{n}{k}$ which is an integer. Two very important examples of divisibility sequences are $a_n = x^n - y^n$ for relatively prime integers x, y and $a_n = F_n$ where F_n is the nth Fibonacci number. In Chap. 4 we will prove a generalization of the result that (F) is a strong divisibility sequence and in Chap. 5 we will generalize even further. $a_n = x^n - y^n$ is the best example to choose because of multiple reasons. First, we readily have $\gcd(x^m - y^m, x^n - y^n) = x^{\gcd(m,n)} - y^{\gcd(m,n)}$. And the second is that we also have the factorization into cyclotomic polynomials

$$x^n - y^n = y^{\varphi(n)} \prod_{d\mid n} \Phi_d\left(\frac{x}{y}\right)$$

3.2 Characterization of Divisibility Sequences

We will start with a basic result that is equivalent to the classical result that the order of a modulo n divides k if $a^k - 1$ is divisible by n.

Theorem 3.1 *If the rank of apparition of m is ρ in a strong divisibility sequence (a), then for any a_k divisible by m, $\rho \mid k$.*

Proof By definition, ρ is the smallest positive integer such that $m \mid a_\rho$. For $m \mid a_k$, assume that $g = \gcd(\rho, k)$ with $0 \le e < \rho$. We have $\gcd(a_k, a_\rho) = a_{\gcd(k,\rho)}$. Since $m \mid a_k$ and $m \mid a_\rho$, $m \mid \gcd(a_k, a_\rho)$. This gives us $m \mid a_g$. Since $g \mid \rho$, we have $g \le \rho$ and $m \mid a_g$. It is not possible to have $g < \rho$ because that would give us a smaller index than ρ for which $m \mid a_g$. Thus, $g = \rho = \gcd(\rho, k)$ implying $\rho \mid k$. ∎

Theorem 3.2 *Let n be a positive integer and $n = p_1^{e_1} \cdots p_k^{e_k}$ be its prime factorization. If the rank of apparition of n in a strong divisibility sequence (a) is ρ, then $\rho = \mathrm{lcm}(\rho_{1_{e_1}}, \rho_{2_{e_2}}, \ldots, \rho_{k_{e_k}})$.*

Proof By definition, $p_i^{e_i} \mid a_{\rho_{i_{e_i}}}$ and since $p_i^{e_i} \mid n$, $p_i^{e_i} \mid a_\rho$. Therefore, $\rho_{i_{e_i}} \mid \rho$ for $1 \le i \le k$. Evidently, $\mathrm{lcm}(\rho_{1_{e_1}}, \ldots, \rho_{k_{e_k}}) \mid \rho$. Since the choice of ρ is minimal, we cannot have $\rho > \mathrm{lcm}(\rho_{1_{e_1}}, \ldots, \rho_{k_{e_k}})$ implying $\rho = \mathrm{lcm}(\rho_{1_{e_1}}, \ldots, \rho_{k_{e_k}})$. ∎

Theorem 3.3 *A divisibility sequence (a) is a strong divisibility sequence if and only if for any prime p and positive integer α, $p^\alpha \mid a_k$ if and only if $\rho_\alpha \mid k$.*

Proof If (a) is a strong divisibility sequence, then $\rho_\alpha \mid k$ if $p^\alpha \mid a_k$ by Theorem 3.1.
   Now, assume that for any prime p and positive integer α, $\rho_\alpha \mid k$ if $p^\alpha \mid a_k$. We want to show that $\gcd(a_m, a_n) = a_{\gcd(m,n)}$. Consider a prime p that divides both a_m and a_n. Let $p^r \parallel a_m$, $p^s \parallel a_n$, $g = \gcd(m, n)$ and $l = \min(r, s)$. We have $p^l \parallel \gcd(a_m, a_n)$. Without loss of generality, $r \le s$ so $l = r$ and $\rho_r \mid m$, $\rho_s \mid n$. Since $p^r \mid a_n$ as well, $\rho_r \mid n$ holds true. Thus, $\rho_r \mid \gcd(m, n)$ and $p^r \parallel a_g$. $p^{r+1} \nmid a_g$ because then $\rho_{r+1} \mid$ and $g \mid m$ which contradicts the fact that $p^r \parallel a_m$. This way, we have $p^r \parallel a_{\gcd(m,n)}$ for any prime p dividing a_m, a_n. ∎

Ward [War55] proved the following theorem that is crucial to characterizing divisibility sequences. It also appeared at Iran mathematical Olympiad 2001, problem 5.

Theorem 3.4 *If (a) is a strong divisibility sequence, then there exists a sequence (b) such that*

$$a_n = \prod_{d \mid n} b_d$$

for every $n \in \mathbb{N}$.

Proof We can assume $n > 1$. Let $\mathrm{rad}(n)$ denote the product of the distinct prime divisors of n. So, $\mathrm{rad}(n)$ is the largest square-free divisor of n. Let $E(n)$ and $O(n)$

be the product of all $a_{\frac{n}{d}}$ such that d is square-free and has an even and odd number of prime factors, respectively. Set $b_n = \frac{E(n)}{O(n)}$.

First, we prove that b_n is an integer. Call a divisor d of n a *top (bottom) divisor* if $\frac{n}{d}$ is square-free and has even (odd) number of prime factors. So, $E(n)$ is the product of a_d for all all top divisors and $O(n)$ is the product of a_d for all bottom divisors. If p is a divisor of (a) and ρ is the rank of apparition, then $\rho \mid d$ for any $p \mid a_d$. We have to show that the number of top divisors d such that $\rho \mid \frac{n}{d}$ is at least the number of bottom divisors d such that $\rho \mid \frac{n}{d}$. The claim is trivially true if $\rho \nmid n$. Now, $\rho \mid \frac{n}{d}$ holds if $d \mid \frac{n}{\rho}$. If $\frac{n}{\rho}$ has l different prime factors, the number of top divisors is $\binom{l}{0} + \binom{l}{2} + \cdots +$ whereas the number of bottom divisors is $\binom{l}{1} + \binom{l}{3} + \cdots$. Since $n > 1$, we have $l = \mathrm{rad}(n) \geq 1$, so both of them equal to 2^{l-1}. We have the conclusion.

Now, we should prove that $a_n = \prod_{d \mid n} b_d$ for b_n defined above.

$$\prod_{d \mid n} b_d = \frac{\prod_{d \mid n} E(d)}{\prod_{d \mid n} O(d)} \tag{3.1}$$

In the numerator of (3.1), $E(d)$ is the product of a_k such that $k \mid d$. Since $d \mid n$, we can say $E(d)$ is the product of a_k such that $k \mid n$. If a_k is part of $E(d)$, then letting $q = \frac{d}{k}$, q is square-free by the definition of $E(d)$ and q has an even number of prime factors. $d \mid n$ also implies that $t \mid \frac{n}{k}$. Conversely, if q is a square-free positive integer such that $q \mid d$ and has an even number of prime factors, then for $k = \frac{d}{q}$, a_k appears in $E(d)$.

If $\frac{n}{q}$ has r distinct prime factors, the number of q is

$$\binom{r}{0} + \binom{r}{2} + \binom{r}{4} + \cdots$$

By a similar argument, the number of times a_q appears in O_d is

$$\binom{r}{1} + \binom{r}{3} + \cdots$$

For $m < n$, the terms above are equal. For $m = n$, we do not have any such prime factors so $l = 0$. But in this case, a_q appears once in the numerator and does not appear in the denominator at all. This means that every other a_d cancel each other out and we only get a_n. So, $\prod_{d \mid n} b_d = a_n$, as desired.   ∎

In fact, we can smooth out Theorem 3.4 further. We can even characterize the sequence (b) itself in more than one ways. For example, we can simply use Möbius inversion (see Möbius Function and Möbius Inversion in glossary) and get

$$b_n = \prod_{d \mid n} a_{\frac{n}{d}}^{\mu(d)}$$

We should mention that a sequence (a) may be written as $a_n = \prod_{d|n} b_d$ even if (a) is not strong divisible. To be more specific, $a_n = \prod_{d|n} b_d$ does not guarantee that (a) is a divisibility sequence. The terms of (b) must maintain some certain properties as well. For example, consider the sequence $1, 1, 2, 2, 2, \ldots$ where the rest of the terms are 2. This is a divisibility sequence but clearly not a strong one. On the other hand, we can check the lcm sequence of (a) with $a_n = n$.

$$n = \prod_{d|n} b_d$$

Clearly, $a_1 = b_1 = 1$. If $n = p$ for a prime p, then $p = b_1 b_p$ so $b_p = p$. Using

$$p^s = \prod_{d|p^s} b_d$$
$$= \prod_{i=0}^{s} b_{p^i}$$

we find $b_{p^i} = p$ for $i \geq 1$. Again,

$$pq = \prod_{d|pq} b_d$$
$$= b_1 b_p b_q b_{pq}$$

gives $b_{pq} = 1$. In a similar manner, we get

$$b_n = \begin{cases} p & \text{if } n = p^s \text{ for some positive integer } s \\ 1 & \text{otherwise} \end{cases}$$

We are interested in finding properties of b_n rather than just the form one shown here.

Theorem 3.5 *For a strong divisibility sequence* (a) *with* $a_n = \prod_{d|n} b_d$, $\gcd(b_m, b_n) = 1$ *if* $m \nmid n$ *and* $n \nmid m$.

Proof Consider m and n such that $\gcd(m, n) = 1$.

$$\gcd(a_m, a_n) = a_{\gcd(m,n)}$$

$$\gcd\left(\prod_{d|m} b_d, \prod_{d|n} b_d \right) = 1$$

This shows that $\gcd(b_d, b_e) = 1$ for $d \mid m$ and $e \mid n$. Let $\gcd(m, n) = g$.

$$\gcd(a_m, a_n) = a_g$$

$$\implies \gcd\left(\prod_{d|m} b_d, \prod_{d|n} b_d\right) = \prod_{d|g} b_g$$

$$\implies \left(\prod_{d|g} b_d\right) \cdot \gcd\left(\prod_{\substack{d \nmid g \\ d|m}} b_d, \prod_{\substack{d \nmid g \\ d|n}} b_d\right) = \prod_{d|g} b_d$$

$$\implies \gcd\left(\prod_{\substack{d \nmid g \\ d|m}} b_d, \prod_{\substack{d \nmid g \\ d|n}} b_d\right) = 1$$

The last equation tells us if $d \mid m$ but $d \nmid n$, and $e \mid n$ but $e \nmid m$, then $\gcd(b_d, b_e) = 1$. In other words, if $m \nmid n$ and $n \nmid m$, then $\gcd(b_m, b_n) = 1$. ∎

In fact, we have the following.

Theorem 3.6 *If (b) is a sequence associated with a strong divisibility sequence (a) such that* $a_n = \prod_{d|n} b_d$, *then (a) is a strong divisibility sequence if and only if* $\gcd(b_m, b_n) = 1$ *whenever* $n \nmid m$ *and* $m \nmid n$.

Theorem 3.7 *For a strong divisibility sequence (a), if*

$$a_n = \prod_{d|n} b_d$$

then $\mathrm{lcm}(a_1, a_2, \ldots, a_n) = b_1 b_2 \cdots b_n$.

Proof We induct on n. Set $L_0 = 1$ and $L_n = \mathrm{lcm}(L_{n-1}, a_n)$. We have $L_1 = a_1 = b_1$. Assume that

$$L_n = \prod_{i=1}^{n} b_i$$

for $n \leq m$. Then

$$L_{m+1} = \mathrm{lcm}(L_m, a_{m+1})$$

$$= \mathrm{lcm}\left(\prod_{i=1}^{m} b_i, \prod_{d|m+1} b_d\right)$$

We can split both arguments into two parts based on the divisors of $m + 1$.

$$L_{m+1} = \operatorname{lcm}\left(\prod_{\substack{d|m+1 \\ d \le m}} b_d \prod_{\substack{d \nmid m+1 \\ d \le m}} b_d, b_{m+1} \prod_{\substack{d|m+1 \\ d \le m}} b_d \right)$$

$$= \prod_{\substack{d|m+1 \\ d \le m}} b_d \operatorname{lcm}\left(\prod_{\substack{d \nmid m+1 \\ d \le m}} b_d, b_{m+1} \right)$$

$$= b_{m+1} \prod_{\substack{d|m+1}} b_d \prod_{\substack{d \nmid m+1 \\ d \le m}} b_d$$

since $\gcd(b_r, b_s) = 1$ for $r \nmid s$, $s \nmid r$. We have

$$L_{m+1} = b_{m+1} \prod_{i=1}^{m} b_i$$

$$= \prod_{i=1}^{m+1} b_i$$

∎

Recall that for an integer sequence (a), its lcm sequence (c) is defined as

$$c_n = \frac{\operatorname{lcm}(a_1, a_2, \ldots, a_n)}{\operatorname{lcm}(a_1, a_2, \ldots, a_{n-1})}$$

Theorem 3.8 *If (a) is a strong divisibility sequence and*

$$a_n = \prod_{d|n} b_d$$

then (b) is the lcm *sequence associated with (a).*

This follows directly from Theorem 3.7. The next theorem is also easily proven.

Theorem 3.9 *If (a) is a strong divisibility sequence, (b) is the* lcm *sequence of (a), and*

$$n = \prod_{i=1}^{k} p_i^{e_i}$$

is the prime factorization of n, then

$$b_n = \frac{a_n}{\operatorname{lcm}\left(a_{\frac{n}{p_1}}, a_{\frac{n}{p_2}}, \ldots, a_{\frac{n}{p_k}} \right)}$$

Combining Theorems 3.4, 3.5, 3.7, and 3.8, we obtain the following result.

Theorem 3.10 *An integer sequence (a) is a strong divisibility sequence if and only if*

$$a_n = \prod_{d \mid n} b_d$$

for all $n \in \mathbb{N}$, where (b) is the lcm *sequence of (a). Moreover,* $\gcd(b_m, b_n) = 1$ *if $m \nmid n$ and $n \nmid m$.*

Note that this result looks like a generalization of a similar result for cyclotomic polynomials. If we think about the common characteristics of the common divisibility sequences (n), $(x^n - y^n)$, (F_n), we can easily see that they have some common properties. They all follow $\gcd(a_m, a_n) = a_{\gcd(m,n)}$ (well, that is why they are strong divisibility sequence in the first place) and $a_n = \prod_{d \mid n} b_d$ for some integers $b_1, b_2, \ldots$. Also, the product of the first k terms divides the product of any k consecutive terms in the sequence. These properties make a clear argument that they have some things in common and those properties can be generalized. That is why it was very important to observe the factorization of $x^n - 1$ into cyclotomic polynomials.

We also have the following combined theorem.

Theorem 3.11 *For a strong divisibility sequence (a) and any positive integer n,*

$$\prod_{d \mid n} a_{\frac{n}{d}}^{\mu(d)} = \frac{\operatorname{lcm}(a_1, a_2, \ldots, a_n)}{\operatorname{lcm}(a_1, a_2, \ldots, a_{n-1})}$$

$$= \frac{a_n}{\operatorname{lcm}\left(a_{\frac{n}{p_1}}, a_{\frac{n}{p_2}}, \ldots, a_{\frac{n}{p_k}}\right)}$$

where $p_1, p_2, \ldots, p_k$ are the distinct prime factors of n.

In a similar fashion, we can prove the next theorem. This result is due to an observation by Dedekind.

Theorem 3.12 *Let (a) be a divisibility sequence and (b) be its* lcm *sequence. Then*

$$b_n = \frac{a_n \prod_{\substack{p_i, p_j \mid n \\ i \neq j}} a_{\frac{n}{p_i p_j}} \cdots}{\prod_{p_i \mid n} a_{\frac{n}{p_i}} \prod_{\substack{p_i, p_j, p_k \mid n \\ i \neq j \neq k \neq i}} a_{\frac{n}{p_i p_j p_k}} \cdots}$$

The proof follows by a simple application of the principle of inclusion and exclusion. We already mentioned that $a_n = x^n - 1$ is a strong divisibility sequence. We see here that (Φ) is just the lcm sequence of (a). Consequently, we obtain the following expressions for $\Phi_n(x)$.

$$\Phi_n(x) = \prod_{d|n}(x^d - 1)^{\mu(\frac{n}{d})} \tag{3.2}$$

$$= \frac{x^n - 1}{\text{lcm}\left(x^{\frac{n}{p_1}} - 1, \ldots, x^{\frac{n}{p_r}} - 1\right)} \tag{3.3}$$

$$= \frac{\text{lcm}(x - 1, x^2 - 1, \ldots, x^n - 1)}{\text{lcm}(x - 1, x^2 - 1, \ldots, x^{n-1} - 1)} \tag{3.4}$$

3.3 Binomial Coefficients

Next, we will discuss a very interesting result. If you remember, the fundamental problem that sparked the first author's interest in divisibility sequences is the problem that the product of n consecutive integers is divisible by $n!$. We give a generalization of this theorem for divisibility sequences due to Ward [War36, Theorem 2] .

Theorem 3.13 *Every strong divisibility sequence is a binomial sequence.*

Proof by Ward Take a prime divisor p of (u). Since

$$\binom{n}{k}_a = \frac{n!_a}{k!_a(n-k)!_a}$$

we just have to show that for every prime p, the exponent of p is at least 0 in $\binom{n}{k}_a$. In an effort to do so, he claimed that the exponent of p in $n!_a$ is

$$\sum_{i=1}^{\infty} \left\lfloor \frac{n}{\rho(p,i)} \right\rfloor \tag{3.5}$$

where $\rho(p,i)$ is the rank of apparition of p^i in (a). This is the same as proving that for a prime p and a positive integer n,

$$\nu_p(n!) = \sum_{i \geq 1} \left\lfloor \frac{n}{p^i} \right\rfloor$$

Observing that $p^i \mid a_k$ if and only if $\rho(p,i) \mid k$, the claim easily follows.

$$\nu_p\left(\binom{n}{k}_a\right) = \sum_{i \geq 1} \left(\left\lfloor \frac{n}{\rho(p,i)} \right\rfloor - \left\lfloor \frac{k}{\rho(p,i)} \right\rfloor - \left\lfloor \frac{n-k}{\rho(p,i)} \right\rfloor \right)$$

Recall that $\lfloor x + y \rfloor \geq \lfloor x \rfloor + \lfloor y \rfloor$ for any real x, y (see Floor and Ceiling Functions in glossary). Using this,

$$\left\lfloor \frac{n}{\rho(p,i)} \right\rfloor \geq \left\lfloor \frac{k}{\rho(p,i)} \right\rfloor + \left\lfloor \frac{n-k}{\rho(p,i)} \right\rfloor \quad \text{so} \quad \left\lfloor \frac{n}{\rho(p,i)} \right\rfloor - \left\lfloor \frac{k}{\rho(p,i)} \right\rfloor - \left\lfloor \frac{n-k}{\rho(p,i)} \right\rfloor \geq 0.$$

Then, the exponent of p is a non-negative integer for any p.   ∎

Another proof We can write $n!_a$ in terms of b_i:

$$n!_a = \prod_{i=1}^{n} a_i$$

$$= \prod_{i=1}^{n} \prod_{d \mid i} b_d$$

$$= \prod_{i=1}^{n} b_i^{\lfloor \frac{n}{i} \rfloor}$$

For the last equation, see Sum and Product of Arithmetic Functions in glossary. This can be used to express $\binom{n}{k}_a$ in terms of b_i:

$$\binom{n}{k}_a = \frac{n!_a}{k!_a (n-k)!_a}$$

$$= \frac{\prod_{i=1}^{n} b_i^{\lfloor \frac{n}{i} \rfloor}}{\prod_{i=1}^{k} b_i^{\lfloor \frac{k}{i} \rfloor} \prod_{i=1}^{n-k} b_i^{\lfloor \frac{n-k}{i} \rfloor}}$$

$$= \prod_{i=1}^{n} b_i^{\lfloor \frac{n}{i} \rfloor - \lfloor \frac{k}{i} \rfloor - \lfloor \frac{n-k}{i} \rfloor}$$

Again, $\lfloor \frac{n}{i} \rfloor \geq \lfloor \frac{k}{i} \rfloor + \lfloor \frac{n-k}{i} \rfloor$ or $\lfloor \frac{n}{i} \rfloor - \lfloor \frac{k}{i} \rfloor - \lfloor \frac{n-k}{i} \rfloor \geq 0$. So, b_i has exponent at least 0 for every i and $\binom{n}{k}_a$ is an integer.   ∎

We can easily generalize this theorem following Carmichael [Car13].

Theorem 3.14 *Let* $\mathbf{m} = \{m_1, m_2, \ldots, m_k\}$ *and* $\mathbf{n} = \{n_1, n_2, \ldots, n_l\}$ *be two sets of positive integers such that for any* $r \geq 1$ *that is a divisor of* r *elements of* $\mathbf{m}$ *is also a divisor of* r *elements of* $\mathbf{n}$. *If* (a) *is a strong divisibility sequence, then*

$$\frac{a_{m_1} a_{m_2} \cdots a_{m_k}}{a_{n_1} a_{n_2} \cdots a_{n_l}}$$

is an integer.

Theorem 3.15 *For any non-negative integers* $n_1, n_2, \ldots, n_k$,

$$\frac{(n_1 + n_2 + \cdots + n_k)!_a}{n_1!_a n_2!_a \cdots n_k!_a}$$

is an integer.

This theorem is analogous to the fact that the multinomial coefficient

$$\frac{(n_1 + n_2 + \cdots + n_k)!}{n_1! n_2! \cdots n_k!}$$

is an integer.

Theorem 3.16 *Let m and n be two relatively prime positive integers. Then for a divisibility sequence* (a),

$$\frac{(m + n - 1)!_a}{m!_a n!_a}$$

is an integer.

Proof Let $d > 1$ be a divisor of k integers from the set $\{1, 2, \ldots, m\}$ and l integers from the set $\{1, 2, \ldots, n\}$. Then d is a divisor of at least $k + l$ integers from the set $\{1, 2, \ldots, m + n - 1\}$. Using Theorem 3.14, we get the result. ∎

We can extend this result even further.

Theorem 3.17 *For a divisibility sequence* (a) *and non-negative integers* $k, m_1,$ $m_2, \ldots, m_r$,

$$\frac{(km_1)!_a (km_2)!_a \cdots (km_r)!_a}{(m_1)!_a^{k-1} (m_2)!_a^{k-1} \cdots (m_r)!_a^{k-1} (m_1 + m_2 + \cdots + m_r)!_a}$$

is an integer where $k \geq 2$.

This is analogous to

$$\frac{(km_1)!(km_2)! \cdots (km_r)!}{m_1! m_2! \cdots m_k!(m_1 + m_2 + \cdots + m_r)!}$$

for $k \geq 2$.

Theorem 3.18 *If* $\gcd(a_{pn}, a_{qn}) = a_n$ *for two distinct primes* p, q *and positive integer* n *in a divisibility sequence* (a), *then for any subsequence* (c) *of* (a) *also has this property* $\gcd(c_{pn}, c_{qn}) = c_n$.

Proof First, consider that p and q are distinct primes. For $c_n = \frac{a_{sn}}{a_s}$,

$$\gcd(c_{pn}, c_{qn}) = \gcd\left(\frac{a_{spn}}{a_s}, \frac{a_{sqn}}{a_s}\right)$$
$$= \frac{\gcd(a_{spn}, a_{sqn})}{a_s}$$
$$= \frac{a_{sn}}{a_s} = c_n$$

If $p = q$, then the claim still holds true because $\frac{\gcd(a_{spn}, a_{spn})}{a_s} = \frac{a_{spn}}{a_s} = c_{pn} = \gcd(c_{pn}, c_{qn})$. ■

Theorem 3.19 *If (a) is a divisibility sequence and $\gcd(a_{pn}, a_{qn}) = a_n$ for any two distinct primes p, q and positive integer n, then $\gcd(a_m, a_n) = 1$ if $\gcd(m, n) = 1$.*

Proof We will use induction using some properties of gcd and lcm (see GCD and LCM in glossary) repeatedly to prove this claim for prime m first. If m and n are just different primes, then the claim is trivially true by setting n to 1. Now assume that $m = p$ is a prime. We must have n not to be a prime. Write $n = q_1 \cdots q_k$ where $q_i \neq p$ are prime factors (possibly distinct) of n and $k > 1$. By induction hypothesis, we can assume that $\gcd(a_p, a_{q_1 \cdots q_i}) = 1$ for $1 \leq i < k$. Setting n to $\frac{n}{q_k}$, q to q_k and p to p, we get $\gcd\left(a_{p\frac{n}{q_k}}, a_n\right) = a_{\frac{n}{q_k}}$.

$$\gcd\left(a_p, a_{\frac{n}{q_k}}\right) = 1$$
$$\gcd\left(a_p, \gcd\left(a_{p\frac{n}{q_k}}, a_n\right)\right) = 1$$
$$\gcd\left(\gcd\left(a_p, a_{p\frac{n}{q_k}}\right), a_n\right) = 1$$
$$\gcd\left(a_p, a_n\right) = 1$$

The last line is true because $p \mid p\frac{n}{q_k}$ so $a_p \mid a_{p\frac{n}{q_k}}$ implying $\gcd\left(a_p, a_{p\frac{n}{q_k}}\right) = a_p$. Thus, the claim is true for prime m. Next, we take $m = p_1 \cdots p_l$ where $p_i \neq q_j$ for $1 \leq i \leq l, 1 \leq j \leq k$. We again use induction. Assume that $\gcd(a_{p_1 \cdots p_i}, a_n) = 1$ for $1 \leq i < l$. Setting n to $\frac{mn}{p_l q_1}$, q to q_1 and p to p_l, we get $\gcd\left(a_{\frac{mn}{q_1}}, a_{\frac{mn}{p_l}}\right) = a_{\frac{mn}{p_l q_1}}$.

$$\gcd\left(a_m, a_{\frac{mn}{p_l q_1}}\right) = \gcd\left(a_m, \gcd\left(a_{\frac{mn}{q_1}}, a_{\frac{mn}{p_l}}\right)\right)$$
$$= \gcd\left(\gcd\left(a_m, a_{\frac{mn}{q_1}}\right), a_{\frac{mn}{p_l}}\right)$$
$$= \gcd\left(a_m, a_{\frac{mn}{p_l}}\right)$$

Again, here $m \mid \frac{mn}{q_1}$ since $q_1 \mid n$ so $a_m \mid a_{\frac{mn}{q_1}}$ implies $\gcd(a_m, a_{\frac{mn}{q_1}}) = a_m$. We repeat this process replacing n with $\frac{n}{q_1}, \frac{n}{q_1 q_2}, \ldots, \frac{n}{q_1 q_2 \cdots q_k}$ and keeping m, p_l unchanged.

$$\gcd\left(a_m, a_{\frac{mn}{p_l}}\right) = \gcd\left(a_m, a_{\frac{mn}{p_l q_1}}\right)$$
$$= \gcd\left(a_m, a_{\frac{mn}{p_l q_1 q_2}}\right)$$
$$= \cdots$$
$$= \gcd\left(a_m, a_{\frac{m}{p_l}}\right)$$
$$= a_{\frac{m}{p_l}}$$

By induction hypothesis, we know

$$\gcd\left(a_{\frac{m}{p_l}}, a_n\right) = 1$$

$$\gcd\left(\gcd\left(a_m, a_{\frac{mn}{p_l}}\right), a_n\right) = 1$$

$$\gcd\left(a_m, \gcd\left(a_{\frac{mn}{p_l}, a_n}\right)\right) = 1$$

$$\gcd\left(a_m, a_n\right) = 1$$

since $n \mid \frac{mn}{p_l}$. Hence, the lemma is true.   ∎

Theorem 3.20 *A necessary and sufficient condition that a divisibility sequence* (*a*) *is a strong divisibility sequence is that* $\gcd(a_{np}, a_{nq}) = a_n$ *for distinct primes* p, q.

Proof We can use Theorems 3.20 and 3.19. Let $g = \gcd(m, n)$ and $m = gu, n = gv$ with $\gcd(u, v) = 1$. Consider the subsequence of (*a*) for $s = g$. Then $c_n = \frac{a_{gn}}{a_g}$ has the property $\gcd(c_{pm}, c_{qm}) = 1$. By the theorem above,

$$\gcd(c_u, c_v) = 1$$

$$\gcd\left(\frac{a_{gu}}{a_g}, \frac{a_{gv}}{a_g}\right) = 1$$

This proves the result.   ∎

References

[Car13]   R.D. Carmichael, On the numerical factors of the arithmetic forms $\alpha^n \pm \beta^n$. Ann. Math. **15**(1/4), 30–48 (1913). ISSN: 0003486X, http://www.jstor.org/stable/1967797

[Hal36]   M. Hall, Divisibility sequences of third order. Amer. J. Math. **58**(3), 577–584 (1936). ISSN: 00029327, 10806377, http://www.jstor.org/stable/2370976

[Leh30]   D.H. Lehmer, An Extended Theory of Lucas' Functions. Ann. Math. **31**(3), 419–448 (1930). ISSN: 0003486X, http://www.jstor.org/stable/1968235

[Luc78a] E. Lucas, Théorie des Fonctions Numériques Simplement Périodiques. Amer. J. Math. **1**(2), 184–196 (1878). ISSN: 00029327, 10806377, https://www.jstor.org/stable/2369308

[Luc78b] E. Lucas, Théorie des Fonctions Numériques Simplement Périodiques. Amer. J. Math. **1**(3), 197–240 (1878). ISSN: 00029327, 10806377, http://www.jstor.org/stable/2369311

[Luc78c] E. Lucas, Théorie des Fonctions Numériques Simplement Périodiques. Amer. J. Math. **1**(4), 289–321 (1878). ISSN: 00029327, 10806377, http://www.jstor.org/stable/2369373

[Now15]   A. Nowicki, Strong divisibility and LCM-sequences. Amer. Math. Monthly **122**(10), 958 (2015). https://doi.org/10.4169/amer.math.monthly.122.10.958

[War36]   M. Ward, Note on divisibility sequences. Bull. Amer. Math. Soc. **42**(12), 843–845 (1936), https://projecteuclid.org:443/euclid.bams/1183499449

[War55]   M. Ward, The mappings of the positive integers into themselves which preserve division. Pacific J. Math. **5**(2), 1013–1023 (1955), https://projecteuclid.org:443/euclid.pjm/1172000966

Chapter 4
Lucas Sequences

This chapter discusses generalizations on Lucas sequence. We will establish some results regarding general Lucas sequences and find out when a Lucas sequence is divisible. The usual Fibonacci sequence $1, 1, 2, 3, 5, 8, \ldots$ is a special case of Lucas sequence. Therefore, pretty much every theorem discussed in this chapter along with the results in Chap. 3 are applicable to Fibonacci numbers.

4.1 Introduction

A sequence $a_1, a_2, \ldots, a_n, \ldots$ is denoted by (a). Fibonacci sequence (F) is defined as $F_0 = 0$, $F_1 = 1$ and $F_n = F_{n-1} + F_{n-2}$ for $n > 1$. *General Fibonacci sequence* is defined as $F_0 = u$, $F_1 = v$, and $F_n = F_{n-1} + F_{n-2}$ for $n > 1$. We define Lucas sequences (u) and (v) the same way Lucas did in [Luc78a]. Let α and β denote the solutions of the equation $x^2 = ax - b$ where a and b are positive or negative relatively prime integers. We have $\alpha + \beta = a$, $\alpha\beta = b$. The difference $\alpha - \beta = \delta$ and $\Delta = a^2 - 4b = \delta^2$. Then

$$U_n = \epsilon \frac{\alpha^n - \beta^n}{\alpha - \beta}$$

$$V_n = \epsilon(\alpha^n + \beta^n)$$

are *Lucas sequences* where $\epsilon \in \{1, -1\}$. We have introduced the term ϵ in order to take absolute values in consideration. The properties remain same regardless of the sign. Therefore, without loss of generality we can simply consider $\epsilon = 1$ and ignore the sign in future. (U) is called a *Lucas sequence of the first kind* and (V) a *Lucas sequence of the second kind*. The *General Lucas* sequence (L) is defined as $L_0 = u$, $L_1 = v$, and $L_n = aL_{n-1} - bL_{n-2}$ for $n > 1$. For now we do not show formulas of L_n like (U) and (V). We will derive those formulas in this chapter. Notice that the formulas above resemble *Binet's formula* for Fibonacci numbers. Also, we usually consider $\alpha \neq \beta$ only or $\delta \neq 0$.

4.2 Divisibility of Second-Order Lucas Sequences

For a general Lucas sequence $L_0 = u$, $L_1 = v$, and $L_n = aL_{n-1} - bL_{n-2}$, we are going to find the starting values u, v and coefficients a, b of (L) for which (L) is a divisibility sequence. When Lucas [Luc78a, Luc78b, Luc78c] defined sequences (U), (V) and investigated their properties, he did not consider these sequences from a more general point of view. He defined (U) as $\dfrac{\alpha^n - \beta^n}{\alpha - \beta}$ right from the start. In this chapter, we will explore if there are sequences similar to (U) and (V) which Lucas could have considered as well. We will show later that, in this aspect, it suffices to consider the sequences Lucas defined himself if we disregard the sign. For now, we will not discuss the basic properties of Lucas sequences because in Chap. 5, we will discuss Lehmer sequences in detail which are a generalization of Lucas sequences.

Theorem 4.1 *A general Lucas sequence (L) is a divisibility sequence if and only if*

$$(u, v, a, b) \in \{(u, v, a, 0), (0, \pm v, a, b)\}$$

Moreover, such a divisibility sequence is also a strong divisibility sequence hence also a binomial sequence.

The theorem asserts that (U) is the only divisibility sequence with no non-trivial divisors.

Theorem 4.2 *Let (L) be a Lucas sequence. $\gcd(L_n, b) = \gcd(b, v)$ for $n > 1$.*

Proof Assume that $\gcd(b, v) = g$ and $b = gc$ and $v = gw$ with $\gcd(c, w) = 1$. For $n > 1$,

$$\begin{aligned}
\gcd(L_n, b) &= \gcd(aL_{n-1} - bL_{n-2}, b) \\
&= \gcd(aL_{n-1}, b) \\
&= \gcd(L_{n-1}, b) \\
&= \cdots \\
&= \gcd(L_1, b)
\end{aligned}$$

This is true because $\gcd(a, b) = 1$. Therefore, $g \mid L_n$ for all $n > 1$.   ■

If $L_n = gL_n'$ for $n > 1$, then we consider the sequence $L_0' = u$, $L_1' = w$ and $L_n' = aL_{n-1}' - bL_{n-2}'$. For simplicity, we replace L' with L. That is, from now $L_0 = u$, $L_1 = v$ and $L_n = aL_{n-1} - bL_{n-2}$ such that $\gcd(b, v) = \gcd(a, b) = 1$. Assuming this, we have $\gcd(L_n, b) = 1$ for all $n > 1$.

Theorem 4.3 *For $n > 1$, $\gcd(L_n, L_{n+1}) = 1$ if $\gcd(u, v) = 1$.*

Proof We employ the Euclidean algorithm again along with the theorem above.

$$
\begin{aligned}
\gcd(L_n, L_{n-1}) &= \gcd(aL_{n-1} - bL_{n-2}, L_{n-1}) \\
&= \gcd(bL_{n-2}, L_{n-1}) \\
&= \gcd(L_{n-2}, L_{n-1}) \\
&= \cdots \\
&= \gcd(L_2, L_1) \\
&= \gcd(v, u)
\end{aligned}
$$

Thus, consecutive terms in Lucas sequence are relatively prime if $\gcd(u, v) = 1$. If $\gcd(u, v) = g > 1$, then we can simply divide (L) by g since L_i is divisible by g for all $i \geq 0$. ∎

Due to this theorem, we can also impose the condition that $\gcd(u, v) = g = 1$. Otherwise, g would be a trivial divisor of (L). Thus, we will assume from now that $L_0 = u$, $L_1 = v$ and $L_n = aL_{n-1} - bL_{n-2}$ with $\gcd(u, v) = \gcd(b, v) = \gcd(a, b) = 1$. So, we will find general Lucas sequences which are divisibility sequences under these conditions.

Theorem 4.4 *If α and β are roots of the equation $x^2 - ax + b = 0$, then for any n,*

$$
L_n = v\frac{\alpha^n - \beta^n}{\alpha - \beta} - ub\frac{\alpha^{n-1} - \beta^{n-1}}{\alpha - \beta}
$$

if $\alpha \neq \beta$, otherwise,

$$
L_n = (vn + u\alpha(1 - n))\alpha^{n-1}
$$

Proof Taking $L_n = \lambda^n$ gives us the characteristic equation of (L), $\lambda^n = a\lambda^{n-1} - b\lambda^{n-2}$ giving us $\lambda^2 - a\lambda + b = 0$. It has two solutions α, β and discriminant $\delta = a^2 - 4b$. If $\delta \neq 0$, then we have $L_n = r\alpha^n + s\beta^n$, otherwise $L_n = (r + sn)\alpha^n$. In both cases, we can use $L_0 = u$ and $L_1 = v$ to find values of r, s. In the first case, $r + s = u$, $r\alpha + s\beta = v$ which gives us $r = \dfrac{v - u\beta}{\alpha - \beta}$, $s = \dfrac{u\alpha - v}{\alpha - \beta}$. Thus,

$$
\begin{aligned}
L_n &= \frac{v - u\beta}{\alpha - \beta}\alpha^n + \frac{u\alpha - v}{\alpha - \beta}\beta^n \\
&= \frac{v\alpha^n - u\beta\alpha^n + u\alpha\beta^n - v\beta^n}{\alpha - \beta} \\
&= \frac{v(\alpha^n - \beta^n) - u\alpha\beta(\alpha^{n-1} - \beta^{n-1})}{\alpha - \beta} \\
&= v\frac{\alpha^n - \beta^n}{\alpha - \beta} - ub\frac{\alpha^{n-1} - \beta^{n-1}}{\alpha - \beta}
\end{aligned}
$$

$$= vU_n - ubU_{n-1}$$

In the second case, $r = u$, $s = \dfrac{v}{\alpha} - u$.

$$L_n = \left(u + \frac{v - u\alpha}{\alpha} n \right) \alpha^n$$
$$= (vn + (1 - n)u\alpha)\alpha^{n-1}$$

∎

As we can see we obtain (U) from (L) if we set $u = 0$ and $v = 1$. Next, we prove the main theorem of this chapter.

Theorem 4.5 *For any Lucas sequence (L) and Lucas sequence of the first kind (U),*

$$L_{m+n+1} = L_{m+1}U_{n+1} - bL_mU_n \tag{4.1}$$

holds for any $m, n \in \mathbb{N}$.

Theorem 4.5 can be proven using induction or Theorem 4.4, but here we give a proof that uses combinatorial interpretations of L_n.

Proof Consider the coloring of an $(n + 2) \times 1$ rectangle with $n + 2$ squares of 1×1. The first square is the starting square S and then follows $n + 1$ squares $0, 1, \ldots, n$. Assume that can paint square S and 0 together in one of u colors and squares from S to square 1 in one of v colors. The rectangle is to be filled up using type A 1×1 tiles or type B 2×1 tiles. Type A and B tiles can be painted with one of a and $c = -b$ colors respectively. Then we can see that the number of ways to tile is $L_n = aL_{n-1} - bL_{n-2}$ for $n > 1$.

Now, consider the tiling starting from square m to square $m + n$. Since there is no starting square S now, the number of tiling is U_{n+1}. If we want to tile a $(m + n + 1) \times 1$ rectangle starting from S, consider the scenarios after reaching square m.

(a) We can go through the $(m + 1)$th square which can be done in L_{m+1} ways. Then we can continue tiling squares from $m + 1$ to $m + n + 1$ in U_{n+1} ways. This way, we have $L_{m+1}U_{n+1}$ ways.

(b) We can bypass the $(m + 1)$th square, that is, after reaching square m, a tile of type B is used. We can color this tile in c ways and then squares $m + 2$ to $m + n + 1$ can be colored in U_n ways. Since we can reach square m in L_m ways, we have cL_mU_n ways.

Therefore, by double counting, $L_{m+n+1} = L_{m+1}U_{n+1} - bL_mU_n$. ∎

Theorem 4.6 *A Lucas sequence of the first kind is a strong divisibility sequence.*

Proof Take two positive integers m, n such that $n = mq + r$ with $0 \leq r < m$. Setting n to $m(q-1) + r - 1$, $U_{mq+r} = U_{m+1}U_{m(q-1)+r} - bU_m U_{m(q-1)+r-1}$.

$$
\begin{aligned}
\gcd(U_m, U_n) &= \gcd(U_m, U_{mq+r}) \\
&= \gcd(U_m, U_{m+1}U_{m(q-1)+r} - bU_m U_{m(q-1)+r-1}) \\
&= \gcd(U_m, U_{m+1}U_{m(q-1)+r}) \\
&= \gcd(U_m, U_{m(q-1)+r}) \text{ since } \gcd(U_m, U_{m+1}) = 1 \\
&= \gcd(U_m, U_{m(q-2)+r}) \\
&= \cdots \\
&= \gcd(U_m, U_r)
\end{aligned}
$$

This is basically the division step of the Euclidean algorithm of the greatest common divisor. Continuing this, we will eventually reach $U_{\gcd(m,n)}$. ∎

Theorem 4.7 *If $b \neq 0$, (L) is a divisibility sequence if and only if $L_n \mid U_n$ for all $n \in \mathbb{N}$.*

Proof Setting n to $mq - 1$,

$$
\begin{aligned}
L_{m(q+1)} &= L_{m+1}U_{mq} - bL_m U_{mq-1} \\
bL_m U_{mq-1} &= L_{m+1}U_{mq} - L_{m(q+1)}
\end{aligned}
$$

If (L) is a divisibility sequence $L_m \mid L_{mq}$ for all $q \geq 1$. Also, we have

$$
\begin{aligned}
L_m &\mid L_{m+1}U_{mq} - L_{m(q+1)} \\
L_m &\mid L_{m+1}U_{mq}
\end{aligned}
$$

for all $q \geq 1$. Since $\gcd(L_m, L_{m+1}) = 1$, $L_m \mid U_{mq}$. Setting $q = 1$, we have $L_m \mid U_m$.

Conversely, assume that $L_m \mid U_m$ for all $m \in \mathbb{N}$. Then since $U_m \mid U_{mq}$, we have $L_m \mid U_{mq}$. Thus,

$$
\begin{aligned}
L_m &\mid L_{m+1}U_{mq} - bL_m U_{mq-1} \\
L_m &\mid L_{m(q+1)}
\end{aligned}
$$

for all $q \geq 1$. Setting $q = 1, 2, \ldots$, we get $L_m \mid L_{2m}, L_{3m}, \ldots$. So, (L) is a divisibility sequence. ∎

Theorem 4.8 *If (L) is a divisibility sequence, then $\gcd(L_m, U_{mq-1}) = 1$ for any $q \geq 1$.*

Proof Let $\gcd(L_m, U_{mq-1})$ be g. We have $g \mid L_m$ so $g \mid U_m$. Thus, $g \mid \gcd(U_m, U_{mq-1})$ implying $g \mid 1$ since $\gcd(U_m, U_{mq-1}) = U_{\gcd(m,mq-1)} = U_{\gcd(m,1)} = U_1$. ∎

Now, back to our original concern, Theorem 4.1. If $b = 0$, $L_{n+1} = aL_n$ for all $n \in \mathbb{N}$. We get $L_n = a^{n-1}v$.

$$\gcd(L_m, L_n) = \left(a^{m-1}v, a^{n-1}v\right)$$
$$= va^{\min(m-1,n-1)}$$

This is not equal to $L_{\gcd(m,n)} = va^{(m,n)-1}$ for all $m, n \in \mathbb{N}$. Therefore, in this case, (L) is only a divisibility sequence but not strong. If $b \neq 0$, then using the theorems above, $L_m \mid U_m$. Setting m to 0, n to $n - 1$,

$$L_n = vU_n - buU_{n-1}$$

Using this equation, $L_n \mid buU_{n-1}$ for all $n \in \mathbb{N}$ whereas $\gcd(L_n, U_{n-1}) = 1$ and $b \neq 0$. Thus, u must be 0 for the relation to hold. Using $u = 0$, $L_n = vU_n$. Since $L_n \mid U_n$, we can not have $|v| > 1$. If $v = 0$, then the sequence is trivially a zero sequence. Therefore, we have $v \in \{1, -1\}$ which proves Theorem 4.1 and a weaker version of Theorem 4.1.

Theorem 4.9 (L) *is a strong divisibility sequence if and only if* (L) *is the Lucas sequence of the first kind.*

Next, we will try to figure out which general Lucas sequences are odd divisibility sequences.

Theorem 4.10 (L) *is an odd divisibility sequence if and only if* $L_m \mid U_{2m}$.

Proof We prove this in a similar way. Setting n to $2mr - 1$ in (4.1), we get

$$L_{m(2r+1)} = L_{m+1}U_{2mr} - bL_mU_{2mr-1}$$
$$bL_mU_{2mr-1} = L_{m+1}U_{2mr} - L_{m(2r+1)}$$
$$L_m \mid L_{m+1}U_{2mr} - L_{m(2r+1)}$$

By definition, $L_m \mid L_{m(2r+1)}$, so we have $L_m \mid L_{m+1}U_{2mr}$. Since $\gcd(L_m, L_{m+1}) = 1$, $L_m \mid U_{2mr}$ for any r. Setting r to 1, we prove the if part of the theorem. The only if part is trivial once we use $L_m \mid U_{2mr}$.   ∎

Theorem 4.11 *If* (L) *is an odd divisibility sequence, then* $\gcd(L_m, U_{2mr-1}) = 1$.

Proof Let $g = \gcd(L_m, U_{2mr-1})$. Then $g \mid L_m$ so $g \mid L_{2mr}$ and $g \mid U_{2mr}$. Since $g \mid U_{2mr-1}$, we have $g \mid \gcd(U_{2mr}, U_{2mr-1})$ and $\gcd(U_{2mr}, U_{2mr-1}) = U_{\gcd(2mr,2mr-1)} = U_1$. Thus, $g = 1$.   ∎

Theorem 4.12 *If* (L) *is an odd divisibility sequence with no trivial divisors, then* $v = 1$.

Proof First, we show that $v \mid L_n$ for all odd n. Clearly, $L_1 = v$ divides $U_2 = a$ implies that $v \mid a$. Now, we can use induction to prove that $v \mid L_{n+2}$ for odd n. $L_{n+1} = aL_{n+1} - bL_n$. Since $v \mid L_n$ and $v \mid a$, we have $v \mid L_{n+2}$. Hence, conclusion.   ∎

Since (L) is a divisibility sequence, we already have that the binomial coefficients of (L) are integers as well. Carmichael [Car13] showed a similar proof. For the fixed real numbers α, β

$$\Phi_n(\alpha, \beta) = \beta^{\varphi(n)} \Phi_n \left(\frac{\alpha}{\beta} \right)$$

We can use Φ_n instead of $\Phi_n(\alpha, \beta)$ since α, β are well defined. Then using the fact that $U_n = \frac{\alpha^n - \beta^n}{\alpha - \beta}$, we get

$$U_n = \prod_{d|n} \Phi_n (\alpha, \beta)$$

using cyclotomic polynomials. This is the same as $a_n = \prod_{d|n} b_d$, so the proof is same as well.

Theorem 4.13 (U) *is a binomial sequence.*

4.3   Lucas Sequence of Arbitrary Order

Inspired by Theorem 4.1, we will consider a Lucas sequence of order k. Consider the integer recurrence (U) defined as

$$U_{n+k} = a_{k-1} U_{n+k-1} + \cdots + a_1 U_{n+1} + a_0 U_n$$

where a_i is an integer for $0 \le i < k$. The characteristic polynomial of this recurrence is

$$f(x) = x^k - a_{k-1} x^{k-1} - \cdots - a_1 x - a_0 \tag{4.2}$$

Let us consider a positive integer $m > 1$ and assume that $\gcd(a_0, m) = 1$. If you remember we defined the period of a Lucas sequence in introduction. We will refine the idea here. Let us call a positive integer ρ *period* of (U) if

$$U_\rho \equiv 0 \pmod{m}$$
$$U_{\rho+1} \equiv 0 \pmod{m}$$

$$\vdots$$

$$U_{\rho+k-2} \equiv 0 \pmod{m}$$
$$U_{\rho+k-1} \equiv 1 \pmod{m}$$

Notice that we have $U_{n+\rho} \equiv U_n \pmod{m}$ for all $n \geq n_0$. If τ is a positive integer such that

$$U_\tau \equiv 0 \pmod{m}$$
$$U_{\tau+1} \equiv 0 \pmod{m}$$
$$\vdots$$
$$U_{\tau+k-2} \equiv 0 \pmod{m}$$

then we call τ the *restricted period* of (U). In other words, τ is a restricted period modulo m if $U_{n+\tau} \equiv aU_n \pmod{m}$ for some integer $a \neq 0$ for all $n \geq n_0$. We again go back to the idea of companion matrix. The companion matrix C of (U) is

$$C = \begin{pmatrix} 0 & 0 & \cdots & 0 & a_0 \\ 1 & 0 & \cdots & 0 & a_1 \\ 0 & 1 & \cdots & 0 & a_2 \\ 0 & 0 & \cdots & 1 & a_{k-1} \end{pmatrix}$$

Like we already observed in Chap. 2, if s_n is the nth state vector $(s_n, s_{n+1}, \ldots, s_{n+k-1})$, then $s_n = s_0 C^n$. Note that, $\det(C) = (-1)^{k-1}a_0$. In order to bear the context of m, we can denote the period and restricted period of (U) by $\rho(m)$ and $\tau(m)$, respectively. But if the context is clear, we will only write ρ and τ. Since $\gcd(m, a_0) = 1$, by Euler's theorem $a_0^{\varphi(m)} \equiv 1 \pmod{m}$. So a_0 is a unit element of the multiplicative group modulo m which is $\{i \in \mathbb{N} : 1 \leq i \leq m, \gcd(i, m) = 1\}$. For two positive integers a and m, let $\mathrm{ord}_n(a)$ be the smallest positive d integer such that $a^d \equiv 1 \pmod{n}$.

Theorem 4.14 *Let p be a prime and (U) be a linear divisibility sequence such that $U_0 = 0, U_1 = 1$. If $p \mid U_n$ for some positive integer n, and τ is the restricted period of (U) modulo p, then $n \mid \tau$.*

Proof First, note that $U_\tau \equiv aU_0 \pmod{p}$, we have $U_\tau \equiv 0 \pmod{p}$. Let d be the greatest common divisor of n and τ and $n = \tau q + r$ with $0 \leq r < \tau$. Since $p \mid U_n$, we also have $p \mid U_{nx}$. Using $U_{n+\tau} \equiv bU_n \pmod{p}$,

$$U_n \equiv U_{q\tau+r} \pmod{p}$$
$$\equiv a^q U_r \pmod{p}$$
$$b^q U_r \equiv 0 \pmod{p}$$

Since $a \not\equiv 0 \pmod{p}$, we have $p \mid U_r$, but $r < \tau$. Thus we must have $r = 0$ and $\tau \mid n$.   ∎

The next couple of theorems are due to Robinson [Rob66].

Theorem 4.15 *Let (U) be a Lucas sequence of order k and m be a positive integer such that $\gcd(m, a_0) = 1$ where a_0 is the coefficient of the least degree in the* recurrence. Let ρ, τ and d be the period, restricted period and $\mathrm{ord}_m((-1)^{k-1}a_0)$, respectively. Then $U_{\tau+k-1}$ is a unit modulo m with order $\frac{\rho}{\tau}$. Moreover, $\frac{\rho}{\tau} \mid kd$.

Proof First, we see that $\rho = \mathrm{ord}(C)$ where $C \in \mathbb{F}_m$. Next, we have $C^n \equiv rI_k$ (mod m) if and only if $U_n \equiv rU_0$ (mod m). So, the restricted period τ creates an ideal of all n such that A^n is a scalar matrix modulo m. Since the period of C is ρ and the order of this ideal is τ, we have $\tau \mid \rho$. Moreover, using the initial state vector $s_0 = (0, 0, \ldots, 0, 1)$,

$$s_n \equiv ts_0 \pmod{m}$$
$$t \equiv U_{n+k-1} \pmod{m}$$
$$C^\tau \equiv U_{\tau+k-1}I_k \pmod{m}$$
$$\det(C^\tau) \equiv U_{\tau+k-1}^k \pmod{m}$$

So, $U_{\tau+k-1}$ is a unit modulo m and $\mathrm{ord}_m(U_{\tau+k-1}) = \frac{\rho}{\tau}$. Since $d = \mathrm{ord}_m((-1)^{k-1}a_0)$, we have $\left((-1)^{k-1}a_0\right)^d \equiv 1 \pmod{m}$.

$$(a_0)^{\tau d} \equiv 1 \pmod{m}$$
$$\left(a_0^\tau\right)^d \equiv 1 \pmod{m}$$
$$U_{\tau+k-1}^{kd} \equiv 1 \pmod{m}$$

Since $\mathrm{ord}_m(U_{\tau+k-1}) = \frac{\rho}{\tau}$, we have $\frac{\rho}{\tau} \mid kd$. ∎

The next theorem is inspired by an investigation into the periodicity of Fibonacci sequence modulo m by Wall [Wal60]. In his own words,

> The most perplexing problem we have met in this study concerns the hypothesis $k(\wp^2) \neq k(\wp)$. We have run a test on a digital computer which shows that $k(\wp^2) \neq k(\wp)$ for all $\wp$ up to 10000; however we can not yet prove that $k(\wp^2) = k(\wp)$ is impossible.

Here, $k(\wp)$ is the period of (U) for a prime $\wp$. Mamangakis [Mam61] proved this hypothesis under a mild assumption.

Theorem 4.16 Let $\wp$ be a prime and c be a positive integer such that $\wp \nmid c$. If $c\wp$ occurs in (U), then $k(\wp^2) \neq k(\wp)$.

Then Robinson [Rob66] showed the following result.

Theorem 4.17 Let p be prime such that $p \nmid a_0$ with notations same as above. Then $\rho(p^2) = \rho(p)$ implies $\tau(p^2) = \tau(p^2)$.

Proof Since $\rho(p) = \mathrm{ord}(C)$ in $\mathbb{F}_p$, $C^{\rho(p)} \equiv I_k$ (mod p). So, we have $C^{\rho(p)} \equiv I_k + pA$ for some matrix A over $\mathbb{F}_p$. Then if we raise them to powers, we easily get

$$C^{p^{e-1}\rho(p)} \equiv I \pmod{p^e}$$

Thus, we have $\rho(p^a) \mid p^{e-1}\rho(p)$. Since $\rho(p) \mid \rho(p^e)$, we have that $\frac{\rho(p^e)}{\rho(p)}$ is power of p. In a similar fashion, we get $\frac{\tau(p^e)}{\tau(p)}$ is a power (possibly non-negative) of p. Now, for any a with $p \nmid a$, $p^{\mathrm{ord}_p(a)} \equiv 1 \pmod{p}$ and $a^{p-1} \equiv 1 \pmod{p}$ by Fermat's little theorem. Therefore, $\mathrm{ord}_p(a) \mid p-1$ so we have $\frac{\rho(p)}{\tau(p)} \mid p-1$ from the previous theorem. So, $p \nmid \frac{\rho(p)}{\tau(p)}$.

$$\frac{\rho(p^e)}{\tau(p)} = \frac{\tau(p^e)}{\tau(p)} \cdot \frac{\rho(p^e)}{\tau(p^e)}$$
$$= \frac{\rho(p^e)}{\rho(p)} \cdot \frac{\rho(p)}{\tau(p)}$$

We can see that the theorem holds true. ∎

If $p \nmid a_0$ and $\rho(p^e) = \rho(p)$, then we can also say

$$\mathrm{ord}_p(U_{\tau(p)+k-1}) = \mathrm{ord}_{p^e}(U_{\tau(p)+k-1})$$

So, we can state the following with the same notations as above.

Theorem 4.18 *Let p be a prime such that $p \nmid a_0$ and e be a positive integer such that $\rho(p^e) = \rho(p)$. Then $\tau(p^e) = \tau(p)$ if and only if $\mathrm{ord}_p(U_{\tau(p)+k-1}) = \mathrm{ord}_{p^e}(U_{\tau(p)+k-1})$.*

Theorem 4.19 *Letting $a_0 \in \{\pm 1\}$ and $p \nmid 2k$ for a prime p, we have $\frac{\rho(p^e)}{\rho(p)} = \frac{\tau(p^e)}{\tau(p)}$ for any positive integer e.*

If (U) is a Lucas sequence of order k and also a divisibility sequence, then we call (U) a *linear divisibility sequence*. We will leave the proof of the next theorem to the readers.

Theorem 4.20 *For a positive integer n,*

$$F_n = \frac{\alpha_1^n}{f'(\alpha_1)} + \frac{\alpha_2^n}{f'(\alpha_2)} + \cdots + \frac{\alpha_k^n}{f'(\alpha_k)}$$

where $f'(\alpha)$ is the evaluation of the derivative of the characteristic polynomial $f(x)$ at $x = \alpha$.

The next theorems are due to Ward [War37]. For these theorems, let (U) be a linear divisibility sequence of order k with no trivial divisors (so $U_1 = 1$ without loss of generality), $\ell = \mathrm{lcm}(1, 2, \ldots, k)$ and $\mathcal{D}$ be its discriminant.

Theorem 4.21 *Let p be a prime. If $p \mid U_p$, then either $p \mid \mathcal{D}$ or $p \mid c_0$.*

Proof If $p \nmid c_0$, then using Theorem 2.4, we see that (U) is periodic modulo p. Then, if $p \nmid \mathcal{D}$, by Theorem 2.36, we have that $p^\ell - 1$ is a period of (U). Now, assume that

ρ is the rank of apparition of p in (U). Then $\rho \mid p$ since $p \mid U_p$. Due to the primality of p, we have $\rho \in \{1, p\}$. If $\rho = 1$, then we have $p \mid U_1 = 1$, a contradiction. Again, if $\rho = p$, since $p^\ell - 1$ is a period, we have $p \mid p^\ell - 1$ or $p \mid 1$, again a contradiction. Thus, at least one of $p \mid \mathcal{D}$ or $p \mid c_0$ has to hold true. ∎

We have the following consequences.

Theorem 4.22 *There is an integer d such that for a prime $p \geq d$, $p \nmid U_p$.*

Theorem 4.23 *For a prime p, $p^k(p^\ell - 1)$ is a period of (U) modulo p.*

Theorem 4.24 *For a prime p, the pre-period of (U) modulo p is at most k.*

Theorem 4.25 *Let p be a prime divisor of U_q for a sufficiently large prime q. Then either (U) is a null sequence modulo p or*

$$p^\ell \equiv 1 \pmod{q}$$

Proof Let q be a prime such that $q > \max(k, d)$ where d is a positive integer such that $p \nmid U_p$ for $p \geq d$. Then we can not have $p = q$. If $q \nmid p^\ell - 1$, then $\gcd(q, p^\ell - 1) = 1$ and $\gcd(q, p^n(p^\ell - 1)) = 1$ for any positive integer n. In this case, we can show that $p \mid U_i$ for $i > k$. The reason is, by Bézout's Identity (see Bézout's Identity for Polynomials in glossary), there are integers x, t such that

$$xq + tp^n(p^\ell - 1) = i$$

Set $n \to k$ so that we have $p^k(p^\ell - 1)$, a period of (U). Without loss of generality, assume that $x > 0$ and $t < 0$, then writing $t = y - z$ where $y < z$,

$$xq + yp^k(p^\ell - 1) = i + zp^k(p^\ell - 1)$$

Take modulo p in the equation above. We have the facts that $p \mid U_q$ implies $p \mid U_{xq}$ and $\gamma = p^k(p^\ell - 1)$ is a period of (U) so $p \mid U_\gamma$. Thus, considering the rank of apparition, we easily see that $p \mid U_i$. Thus, $p \mid U_i$ for every $i > k$ and (U) is a null sequence modulo p. ∎

Theorem 4.26 *Then for any sufficiently large prime number p, we have*

$$U_p^\ell \equiv 1 \pmod{p}$$

Proof Let the prime factorization of U_p be

$$U_p = p_1^{e_1} p_2^{e_2} \cdots p_r^{e_r}$$

By definition none of p_i is a trivial divisor of (U). Consider the prime $q > \max(k, d)$. Using the theorem above,

$$p_i^\ell \equiv 1 \pmod q$$
$$p_i^{e_i\ell} \equiv 1 \pmod q$$

Multiplying the congruences for $1 \le i \le r$, we get

$$p_1^{e_1\ell} \cdots p_r^{e_r\ell} \equiv 1 \pmod q$$
$$U_q^\ell \equiv 1 \pmod q$$

∎

Theorem 4.27 *If (U) has trivial divisors other than ± 1, then (U) contains an infinite number of subsequence (C) have no trivial divisors so that the characteristic polynomial of such a sequence follows the properties of the polynomial in (4.2).*

Proof A positive integer d is a *proper* trivial divisor of (U) if d does not divide any of the initial terms $U_1, U_2, \ldots, U_{k-1}$ or the coefficients $a_{k-1}, \ldots, a_0$. Without loss of generality, assume that the trivial divisors of (U) are proper, $U_0 = 0$, $U_1 = 1$. Then the trivial divisors of (U) are the divisors of $a_{k-1}, a_{k-2}, \ldots, a_0$. Consider a subsequence (C) of (U) with $C_n = \frac{U_{sn}}{U_s}$ and set $n \to s$ in Theorem 1.35. We see that (C) has $f^s(x)$ as characteristic polynomial so if $\gcd(a_{k-1}, \ldots, a_0, p) = 1$, we also have $\gcd(b_{k-1}, \ldots, b_0, p) = 1$. This implies that (C) can not have any trivial divisors. ∎

Consider the following polynomials.

$$f_0(x) = 0$$
$$f_r(x) = x^r - a_{k-1}x^{k-1} - \cdots - a_1 x - a_0 \text{for } r \ge 1$$
$$U(x) = U_0 f_{k-1}(x) + U_1 f_{k-2}(x) + \cdots + U_{k-1} f_0(x)$$

$U(x)$ is the generator of (U).

$$\det(U) = \begin{vmatrix} U_0 & U_1 & \cdots & U_{k-1} \\ U_1 & U_2 & \cdots & U_k \\ \vdots & \vdots & \vdots & \\ U_{k-1} & U_k & \cdot & U_{2k-2} \end{vmatrix}$$
$$= (-1)^{\frac{k(k-1)}{2}} \mathfrak{R}(U(x), f(x))$$
$$= \beta_1 \beta_2 \cdots \beta_k \mathcal{D}$$

where $\beta_1, \beta_2, \ldots, \beta_k$ are the coefficients such that

$$U_n = \beta_1 \alpha_1^n + \beta_2 \alpha_2^n + \cdots + \beta_k \alpha_k^n$$

with $\alpha_1, \alpha_2, \ldots, \alpha_k$ being the distinct roots of $f(x)$ and $\beta_i = \frac{U(\alpha_i)}{f'(\alpha_i)}$. Let us recall the fundamental theorems on null sequences and periodic sequences along with the theorems regarding the calculation of pre-period and period. We can easily imply the following theorem.

Theorem 4.28 *The necessary and sufficient condition that (U) is a linear divisibility sequence with no trivial divisors is* $\gcd(e_0, e_1, \ldots, e_k) = 1$, *where*

$$e_0 = \gcd(U_0, U_1, \ldots, U_{k-1})$$
$$e_1 = \gcd(a_0, U_1, \ldots, U_{k-1})$$
$$e_2 = \gcd(a_0, a_1, U_2, \ldots, U_{k-1})$$

$$\vdots$$

$$e_{k-1} = \gcd(a_0, a_1, \ldots, U_{k-1})$$
$$e_k = \gcd(a_0, a_1, \ldots, a_{k-1})$$

We can easily show that if p is a trivial prime divisor of (U), then $p \mid e_i$ for some $0 \le i < k$. If we assume (U) only has proper trivial divisors, we have $e_k = 1$. Since the pre-period is at most k, we have $p \mid U_{k-1}$, $p \mid U_k$, and so on. Thus, $p \mid \det(U)$. Using $e_0 = e_1 =$ and $p \mid a_0$, $p \mid a_1$, we see that $f(x) \equiv 0 \pmod{p}$ has double root at $x = 0$. Thus, we also have $p \mid \mathcal{D}$. We have the following theorems.

Theorem 4.29 *Let p be a trivial prime divisor of a linear divisibility sequence (U) such that $U_0 = 0$, $U_1 = 1$. Then p divides both $\det(U)$ and $\mathcal{D}$.*

Theorem 4.30 *A sufficient condition that a linear divisibility sequence (U) has no trivial divisors is* $\gcd(\mathcal{D}, \det(U)) = 1$.

Theorem 4.31 (Carmichael) *If a prime p is a trivial divisor of a linear divisibility sequence, then $p \mid a_0$.*

We can prove it using the fact that if $p \nmid a_0$, then (U) is purely periodic modulo p, which we have already shown. For a trivial prime divisor p of a linear divisibility sequence (U), the largest positive integer d for which p^d is a trivial divisor of (U) is the *index* of p in (U). We have the following theorem.

Theorem 4.32 *Let p be a trivial divisor of a linear divisibility sequence (U). If $\alpha = \nu_p(\det(U))$, then the index of p in (U) is at most α.*

We will use this theorem to prove the following result.

Theorem 4.33 *Any subsequence of a linear divisibility sequence (U) cannot have any trivial prime divisor that is not a trivial divisor of (U) itself.*

Proof A prime p is a trivial divisor of (U) if $p \mid a_0$. Consider a subsequence (C) of (U) with $c_n = \frac{U_{sn}}{U_s}$. We have already shown that if (C) has a trivial divisor, then it is a proper trivial divisor. So, the constant term of the characteristic polynomial of (C) must be divisible by p. But this constant term c_0 will divide a power of a_0, thus $p \mid a_0$ as well. ∎

We again consider the polynomial $f^s(x)$ as in Theorem 1.35.

$$f^s(x) = (x - \alpha_1^s)(x - \alpha_2^s) \cdots (x - \alpha_k^s)$$

where $\alpha_1, \alpha_2, \dots, \alpha_k$ are the roots of $f(x)$. So, $f^s(x)$ is the polynomial whose roots are the sth power of the roots of $f(x)$. Similarly, let $\mathcal{D}_s$ be the discriminant of $f^s(x)$. Then, $\frac{\mathcal{D}_s}{\mathcal{D}}$ is an integer.

Theorem 4.34 *There are infinitely many s such that* $\gcd\left(\frac{\mathcal{D}_s}{\mathcal{D}}, \mathcal{D}\right) = 1$.

Proof Let p be an arbitrary prime divisor of $\mathcal{D}$. Consider the field $\mathbb{F}$ generated by $f(x)$.

$$\mathcal{D} = \prod_{1 \le i < j \le k} (\alpha_i - \alpha_j)^2$$

If $\mathfrak{p}$ is a prime ideal factor of p in $\mathbb{F}$, $p \mid \mathcal{D}$ is possible only if

$$\alpha_i \equiv \alpha_j \quad (\text{mod } \mathfrak{p})$$

for some $1 \le i < j \le k$.

$$\frac{\mathcal{D}_s}{\mathcal{D}} = \prod_{1 \le i < j \le k} \frac{\alpha_i^s - \alpha_j^s}{\alpha_i - \alpha_j}$$

$$\frac{\alpha_i^s - \alpha_j^s}{\alpha_i - \alpha_j} \equiv s \quad (\text{mod } (\alpha_i - \alpha_j))$$

where $(\alpha_i - \alpha_j)$ is the principal ideal generated by $\alpha_i - \alpha_j$. Thus, if $\alpha_i \equiv \alpha_j$ (mod $\mathfrak{p}$), then

$$\frac{\alpha_i^s - \alpha_j^s}{\alpha_i - \alpha_j} \equiv 0 \quad (\text{mod } \mathfrak{p})$$

holds true if and only if $s \equiv 0$ (mod p). Choose a positive integer n such that $p \nmid s$. Then if $\gcd\left(\dfrac{\mathcal{D}_s}{\mathcal{D}}, \mathcal{D}\right) > 1$ and p is a common prime divisor, there must be indexes k and l such that

$$\frac{\alpha_k^s - \alpha_l^s}{\alpha_k - \alpha_l} \equiv 0 \quad (\text{mod } \mathfrak{p})$$

$$\alpha_k \not\equiv \alpha_l \quad (\text{mod } \mathfrak{p})$$

$$\gcd(\alpha_k, \mathfrak{p}) = 1$$

$$\gcd(\alpha_l, \mathfrak{p}) = 1$$

$$\gcd(\alpha_k - \alpha_l, \mathfrak{p}) = 1$$

Let d be the least positive integer such that

$$\alpha_k^d \equiv \alpha_l^d \pmod{\mathfrak{p}}$$

Then for any n for which $\alpha_k^n \equiv \alpha_l^n \pmod{\mathfrak{p}}$, we have $d \mid n$ and also $d \mid p^t - 1$ such that $t \leq k!$. For a prime p, let L_p be the least common multiple

$$L_p = \operatorname{lcm}(p - 1, p^2 - 1, \ldots, p^{k!} - 1)$$

Let L be the least common multiple of L_p for all such possible primes p. We can choose n such that $\gcd(s, L) = \gcd(s, \mathcal{D}) = 1$, we see that $\frac{\mathcal{D}_s}{\mathcal{D}}$ is relatively prime to $\mathcal{D}$. Because, if p is a prime divisor of $\frac{\mathcal{D}_s}{\mathcal{D}}$ and $\gcd(\mathcal{D}, s) = 1$, then $d \mid s$ but $\gcd(s, L) = 1$ and $d \mid L$. That would be possible only if $d = 1$, however this is impossible as well due to $\alpha_l \not\equiv \alpha_k \pmod{\mathfrak{p}}$. ∎

We can point out some things in this proof. First, the factorization is very similar to cyclotomic polynomials. Then we prove a result that is similar to if $\Phi_d(x)$ and $\Phi_n(x)$ have a common prime factor p, then $p \mid n$. Then we make use of something that is analogous to $\operatorname{ord}_n(a)$ divides k if n divides $a^k - 1$. We also use an analogous version of the following: let x and y be relatively prime positive integers, then

$$x^n - y^n = (x - y)(x^{n-1} + x^{n-2}y + \cdots + y^{n-1})$$

$$\frac{x^n - y^n}{x - y} \equiv x^{n-1} + x^{n-2}y + \cdots + y^{n-1} \pmod{x - y}$$

$$\equiv nx^{n-1} \pmod{x - y}$$

$$\gcd\left(\frac{x^n - y^n}{x - y}, x - y\right) = \gcd(nx^{n-1}, x - y)$$

$$= \gcd(n, x - y) \tag{4.3}$$

This is a very crucial result that is used in many forms and a lot of important results rely on this simple observation. We will see some of those cases in Chap. 6.

Let p be a prime that does not divide $\mathcal{D}$. Consider the decomposition of p into prime ideals

$$p = \mathfrak{p}_1 \mathfrak{p}_2 \cdots \mathfrak{p}_l$$

$\mathfrak{p}_1, \mathfrak{p}_2, \ldots$ are distinct and none of them have power higher than 1. Let τ_i be the smallest positive integer n such that

$$\alpha_1^n \equiv \alpha_2^n \equiv \cdots \equiv \alpha_k^n \pmod{\mathfrak{p}_i}$$

If $\gcd(p, \det(U)) = 1$, then restricted period τ is the smallest positive integer n such that

$$\alpha_1^n \equiv \alpha_2^n \equiv \cdots \equiv \alpha_k^n \pmod{p}$$

holds true. Thus, we also get the following.

Theorem 4.35 *Let (U) be a linear divisibility sequence and p be a prime that divides neither $\det(U)$ nor $\mathcal{D}$. Then the restricted period of (U) modulo p is the least common multiple of τ_i where τ_i is defined above.*

If $f(x)$ is irreducible, then $f(x)$ has distinct roots and we can write $f(x)$ as

$$U_n = b_1\alpha_1^n + b_2\alpha_2^n + \cdots + b_k\alpha_k^n$$

for some constant $b_1, b_2, \ldots, b_k$ dependent on the initial state vector $s(U)$. We can express the coefficients in terms of the generator polynomial of (U) and the derivative of the characteristic polynomial.

$$b_i = \frac{U(\alpha_i)}{f'(\alpha_i)}$$

Here, $U(\alpha_i)$ is calculated as usual setting $x \to \alpha_i$ in the generator $U(x)$ of (U), $f'(\alpha_i)$ is calculated similarly.

$$\mathcal{D} = f'(\alpha_1)f'(\alpha_2)\cdots f'(\alpha_k)$$

Theorem 4.36 *Let p be a prime. Then*

$$\gcd(p, \mathcal{D}) = \gcd(p, b_1 b_2 \cdots b_k) = 1$$
$$\Longleftrightarrow \gcd(p, \det(U)) = 1$$

Proof We already have that

$$\det(U) = b_1 b_2 \cdots b_k D$$
$$= U(\alpha_1)U(\alpha_2)\cdots U(\alpha_k)$$

Consider the sum of b_j where i of the b_j are taken at the same time (this can be done in $\binom{k}{i}$ ways)

$$b_1 + b_2 + \cdots + b_i$$
$$b_2 + b_3 + \cdots + b_k b_{i+1}$$
$$\cdots$$
$$b_{k-i+1} + b_{k-i+2} + \cdots b_k$$

Let B_i be the product of the $\binom{k}{i}$ terms.

$$B_i = \prod (b_1 + b_2 + \cdots + b_i)$$

Then we have that $B = B_1 B_2 \cdots B_k$ is a rational number and that the prime divisors of the denominator are prime divisors of $\mathcal{D}$.   ∎

We will use this number B in the next theorem as well.

Theorem 4.37 *There is a rational number $B = \frac{P}{Q}$ dependent on the initial state vector of (U) and the coefficients of the characteristic polynomial $f(x)$. Let p be a prime number such that $\gcd(p, P) = \gcd(p, Q) = 1$. Then the rank of apparition of p in (U) is the restricted period of p in (U).*

Proof Let $\mathfrak{p}$ be a prime ideal factor of p and ρ be the rank of apparition of p in (U). By definition, we have that $p \mid U_{n\rho}$ for any positive integer n.

$$b_1 \alpha_1^{np} + b_2 \alpha_2^{np} + \cdots + b_k \alpha_k^{np} \equiv 0 \pmod{\mathfrak{p}}$$

The determinant of this system of congruence modulo p must be divisible by $\mathfrak{p}$ since the integers $b_1, b_2, \ldots, b_k$ are relatively prime to $\mathfrak{p}$. As in Vandermonde determinant, this determinant is a product of differences $\alpha_i^p - \alpha_j^p$. So, all differences cannot be non-zero modulo p. In fact, we can claim that

$$\alpha_1^p \equiv \alpha_2^p \equiv \cdots \equiv \alpha_k^p \pmod{\mathfrak{p}}$$

If not, then for some $l < k$,

$$c_1 \beta_1^n + c_2 \beta_2^n + \cdots + c_l \beta_l^n \equiv 0 \pmod{\mathfrak{p}}$$

such that c_i is found in one of the $\binom{k}{i}$ sums of $\alpha_1, \alpha_2, \ldots, \alpha_k$ and β_i are distinct modulo $\mathfrak{p}$ with

$$\alpha_{i_j}^p \equiv \beta_i \pmod{\mathfrak{p}}$$

for some $i_1, i_2, \ldots < k$. Similarly, we get a system of linear congruence such that the product of differences β_i is not divisible by $\mathfrak{p}$. Thus, the determinant of that system is not divisible by $\mathfrak{p}$. We get that $p \mid B$ which is a contradiction. On the other hand, in a similar manner to order, we get that $\tau \mid \rho$ where $\tau(\mathfrak{p})$ is the restricted period. Thus, $\tau = \mathrm{lcm}(\tau(\mathfrak{p}))$ for all prime ideal divisor $\mathfrak{p}$ of p and $\tau \mid \rho$. But, since $p \mid \tau$, we have $\tau = \rho$.   ∎

Let us take a prime p and a linear recurrent sequence (U) of order k. If p divides l consecutive terms $U_n, U_{n+1}, \ldots, U_{n+l-1}$ for some n, but not U_{n+l}, then p is called a *prime divisor of order l*. We have that $l \leq k$. If $l = k$, following Ward [War54] we say that p is a *maximal prime divisor* of (U). Then using the impulse sequence of (U) which we mentioned as (F) above, we have the following theorem.

Theorem 4.38 *Let (U) be a linear recurrent sequence and p be a prime that does not divide $\mathcal{D}\det(U)a_0$. Moreover, assume that the characteristic polynomial $f(x)$ of (U) has no multiple roots. If p is a maximal divisor of (U), then the rank of apparition of p in (U) divides the rank of p in (F).*

We will leave the proof to the reader although it could be argued that we have almost proven this already. Under the same condition, we can also prove the following.

Theorem 4.39 *p is a maximal divisor of (U) if*

 (i) *$f(x)$ is irreducible and $\deg(f)$ is odd.*
 (ii) *$\gcd(\deg(f), p-1) = 1$.*
 (iii) *$f(x)$ is irreducible modulo p.*

References

[Car13]   R.D. Carmichael, On the numerical factors of the arithmetic forms $\alpha^n \pm \beta^n$. Ann. Math. **15**(1/4), 30–48 (1913). ISSN: 0003486X, http://www.jstor.org/stable/1967797

[Luc78a]   E. Lucas, Théorie des Fonctions Numériques Simplement Périodiques. Amer. J. Math. **1**(2), 184–196 (1878). ISSN: 00029327, 10806377, https://www.jstor.org/stable/2369308

[Luc78b]   E. Lucas, Théorie des Fonctions Numériques Simplement Périodiques. Amer. J. Math. **1**(3), 197–240 (1878). ISSN: 00029327, 10806377. http://www.jstor.org/stable/236931

[Luc78c]   E. Lucas, Théorie des Fonctions Numériques Simplement Périodiques. Amer. J. Math. **1**(4), 289–321 (1878). ISSN: 00029327, 10806377, http://www.jstor.org/stable/2369373

[Mam61]   S.E. Mamangakis, Remarks on the Fibonacci Series Modulo? Amer. Math. Monthly **68**(7), 648 (1961). https://doi.org/10.2307/2311514

[Rob66]   D.W. Robinson, A note on linear recurrent sequences modulo? Amer. Math. Monthly **73**(6), 619–621 (1966). ISSN: 00029890, 19300972, http://www.jstor.org/stable/2314796

[Wal60]   D.D. Wall, Fibonacci series modulo? Amer. Math. Monthly **67**(6), 525–532 (1960). ISSN: 00029890, 19300972, http://www.jstor.org/stable/2309169

[War37]   M. Ward, Linear divisibility sequences. Trans. Amer. Math. Soc. **41**(2), 276–286 (1937). ISSN: 00029947, http://www.jstor.org/stable/1989623

[War54]   M. Ward, The maximal prime divisors of linear recurrences. Canad. J. Math. **6**, 455–462 (1954). https://doi.org/10.4153/cjm-1954-047-9

Chapter 5
Lehmer Sequences

In this chapter, we discuss a generalization of Lucas sequences due to D. H. Lehmer.

5.1 Introduction

Lehmer [Leh30] defined (U) and (V) as

$$U_n = \frac{\alpha^n - \beta^n}{\alpha - \beta}$$
$$V_n = \alpha^n + \beta^n$$

where α, β are distinct roots of the equation $x^2 - \sqrt{c}x + b = 0$ for some relatively prime positive integers c, b. In the original papers of Lucas [Luc78a, Luc78b, Luc78c] a positive integer a which is relatively prime to b was used in place of $\sqrt{c}$. The sequences (U) and (V) both satisfy a second-order linear recurrence:

$$U_{n+2} = aU_{n+1} - bU_n$$
$$V_{n+2} = aV_{n+1} - bV_n$$

where $a = \sqrt{c}$. Without loss of generality, we can assume that $\gcd(b, c) = 1$, $b \neq 0$, $c > 0$. We will also assume a is real, positive, and α/β is not a root of unity. Using Vieta's formulas,

$$\alpha + \beta = \sqrt{c}$$
$$\alpha\beta = b$$
$$\delta = \alpha - \beta = \sqrt{c - 4b}$$
$$\Delta = \delta^2 = c - 4b$$
$$2\alpha = \sqrt{c} + \delta$$
$$2\beta = \sqrt{c} - \delta$$

We will also assume that $\Delta \neq 0$. Using the following, we can prove by induction that U_n and V_n are integers for odd and even n, respectively.

$$U_{n+2} = \sqrt{c}U_{n+1} - bU_n$$
$$V_{n+2} = \sqrt{c}V_{n+1} - bV_n$$

In order to make them integers for all n we make the following modifications.

$$\bar{U}_n = \begin{cases} \dfrac{\alpha^n - \beta^n}{\alpha - \beta} & \text{if } n \text{ is odd} \\ \dfrac{\alpha^n - \beta^n}{\alpha^2 - \beta^2} & \text{otherwise} \end{cases} \tag{5.1}$$

$$\bar{U}_n = \begin{cases} \dfrac{\alpha^n + \beta^n}{\alpha + \beta} & \text{if } n \text{ is odd} \\ \alpha^n + \beta^n & \text{otherwise} \end{cases} \tag{5.2}$$

Such α, β are called a *Lehmer pair*, while $(\bar{U})$ and $(\bar{V})$ are called the associated *Lehmer sequences*. In the original paper, Lehmer briefly considered this modification.

$$\bar{U}_{n+2} = \begin{cases} c\bar{U}_{n+1} - b\bar{U}_n & \text{if } n \text{ is odd} \\ \bar{U}_{n+1} - b\bar{U}_n & \text{otherwise} \end{cases}$$

$$\bar{V}_{n+2} = \begin{cases} c\bar{V}_{n+1} - b\bar{V}_n & \text{if } n \text{ is even} \\ \bar{V}_{n+1} - b\bar{V}_n & \text{otherwise} \end{cases}$$

Lehmer stated the following identities that hold for Lucas sequences as well.

$$V_n^2 - \Delta U_n^2 = 4b^n \tag{5.3}$$
$$2U_{m+n} = U_m V_n + U_n V_m \tag{5.4}$$
$$2V_{m+n} = V_m V_n + \Delta U_m U_n \tag{5.5}$$
$$2b^m U_{m-n} = U_m V_n - V_n U_m \tag{5.6}$$
$$2b^m V_{m-n} = V_m V_n - \Delta U_m U_n \tag{5.7}$$
$$U_{2n} = U_n V_n \tag{5.8}$$

$$V_{2n}^2 = V_n^2 - 2b^n \tag{5.9}$$

with $m > n$ in (5.6)–(5.7). If k is odd, then we also have

$$U_{nk} = \sum_{i=0}^{\frac{k-1}{2}} \frac{k}{k-i} \binom{k-i}{i} b^{ni} \delta^{k-2i-1} U_n^{k-2i} \tag{5.10}$$

$$= \delta^{k-1} U_n^k + \sum_{i=1}^{\frac{k-1}{2}} \frac{k}{i} \binom{k-i-1}{i-1} \delta^{k-2i-1} U_n^{k-2i} \tag{5.11}$$

These identities are best proved with the expansions of expressions containing α, β. For example,

$$(\alpha^m - \beta^m)(\alpha^n + \beta^n) - (\alpha^n - \beta^n)(\alpha^m + \beta^m) = 2\alpha^n \beta^n (\alpha^{m-n} - \beta^{m-n})$$

gives us the identity $U_m V_n - U_n V_m = 2b^n U_{m-n}$ for $m > n$. Let us check that the identity holds for $(\bar{U})$, $(\bar{V})$. First, consider the case when both m and n are odd.

$$\begin{aligned}
\bar{U}_m \bar{V}_n - \bar{U}_n \bar{V}_m &= \frac{\alpha^m - \beta^m}{\alpha - \beta} \cdot \frac{\alpha^n + \beta^n}{\alpha + \beta} - \frac{\alpha^n - \beta^n}{\alpha - \beta} \cdot \frac{\alpha^m + \beta^m}{\alpha + \beta} \\
&= \frac{(\alpha^m - \beta^m)(\alpha^n + \beta^n)}{\alpha^2 - \beta^2} - \frac{(\alpha^n - \beta^n)(\alpha^m + \beta^m)}{\alpha^2 - \beta^2} \\
&= \frac{2\alpha^n \beta^n (\alpha^{m-n} - \beta^{m-n})}{\alpha^2 - \beta^2}
\end{aligned} \tag{5.12}$$

Since $m - n$ is even, we have

$$\frac{\alpha^{m-n} - \beta^{m-n}}{\alpha^2 - \beta^2} = \bar{U}_{m-n}$$

Therefore,

$$\bar{U}_m \bar{V}_n - \bar{U}_n \bar{V}_m = 2b^n \bar{U}_{m-n}$$

Next, assume that both m and n are even. Then

$$\begin{aligned}
\bar{U}_m \bar{V}_n - \bar{U}_n \bar{V}_m &= \frac{\alpha^m - \beta^m}{\alpha^2 - \beta^2}(\alpha^n + \beta^n) - \frac{\alpha^n - \beta^n}{\alpha^2 - \beta^2}(\alpha^m + \beta^m) \\
&= \frac{(\alpha^m - \beta^m)(\alpha^n + \beta^n)}{\alpha^2 - \beta^2} - \frac{(\alpha^n - \beta^n)(\alpha^m + \beta^m)}{\alpha^2 - \beta^2} \\
&= \frac{2\alpha^n \beta^n (\alpha^{m-n} - \beta^{m-n})}{\alpha^2 - \beta^2} \\
&= 2b^n \bar{U}_{m-n}
\end{aligned}$$

since $m - n$ is even. In both cases, the identities are exactly same. We only have two more cases left, where m and n have different parities. Let us first assume that m is odd and n is even.

$$\bar{U}_m \bar{V}_n - \bar{U}_n \bar{V}_m = \left(\frac{\alpha^m - \beta^m}{\alpha - \beta} \right) \cdot (\alpha^n + \beta^n) - \frac{\alpha^n - \beta^n}{\alpha^2 - \beta^2} (\alpha^m + \beta^m)$$

As we can see, the numerators are not the same. So we instead consider

$$
\begin{aligned}
\bar{U}_m \bar{V}_n - c \bar{U}_n \bar{V}_m &= \left(\frac{\alpha^m - \beta^m}{\alpha - \beta} \right) \cdot (\alpha^n + \beta^n) - (\alpha + \beta)^2 \left(\frac{\alpha^n - \beta^n}{\alpha^2 - \beta^2} \right) \cdot \left(\frac{\alpha^m + \beta^m}{\alpha + \beta} \right) \\
&= \frac{(\alpha^m - \beta^m)(\alpha^n + \beta^n) - (\alpha^n - \beta^n)(\alpha^m + \beta^m)}{\alpha - \beta} \\
&= \frac{2\alpha^n \beta^n (\alpha^{m-n} - \beta^{m-n})}{\alpha - \beta} \\
&= 2 b^n \bar{U}_{m-n}
\end{aligned}
$$

since $m - n$ is odd. Finally, if m is even and n is odd,

$$\bar{U}_m \bar{V}_n - \bar{U}_n \bar{V}_m = \left(\frac{\alpha^m - \beta^m}{\alpha^2 - \beta^2} \right) \cdot \left(\frac{\alpha^n + \beta^n}{\alpha + \beta} \right) - \left(\frac{\alpha^n - \beta^n}{\alpha - \beta} \right) \cdot (\alpha^m + \beta^m)$$

In this case, we make the following modification.

$$
\begin{aligned}
c \bar{U}_m \bar{V}_n - \bar{U}_n \bar{V}_m &= (\alpha + \beta)^2 \left(\frac{\alpha^m - \beta^m}{\alpha^2 - \beta^2} \right) \cdot \left(\frac{\alpha^n + \beta^n}{\alpha + \beta} \right) - \left(\frac{\alpha^n - \beta^n}{\alpha - \beta} \right) \cdot (\alpha^m + \beta^m) \\
&= \frac{(\alpha^m - \beta^m)(\alpha^n + \beta^n)}{\alpha - \beta} - \frac{(\alpha^n - \beta^n)(\alpha^m + \beta^m)}{\alpha - \beta} \\
&= \frac{2\alpha^n \beta^n (\alpha^{m-n} - \beta^{m-n})}{\alpha - \beta} \\
&= 2 b^n \bar{U}_{m-n}
\end{aligned}
$$

since $m - n$ is odd. We will also check the analogous formula for V_n for m and n odd.

$$
\begin{aligned}
c \bar{V}_m \bar{V}_n - \Delta \bar{U}_m \bar{U}_n &= (\alpha + \beta)^2 \left(\frac{\alpha^m + \beta^m}{\alpha + \beta} \right) \cdot \left(\frac{\alpha^n + \beta^n}{\alpha + \beta} \right) - (\alpha - \beta)^2 \left(\frac{\alpha^n - \beta^n}{\alpha - \beta} \right) \cdot \left(\frac{\alpha^m - \beta^m}{\alpha - \beta} \right) \\
&= (\alpha^m + \beta^m)(\alpha^n + \beta^n) - (\alpha^m - \beta^m)(\alpha^n - \beta^n) \\
&= \alpha^{m+n} + \beta^{m+n} + \alpha^m \beta^n + \alpha^n \beta^m - (\alpha^{m+n} + \beta^{m+n} - \alpha^m \beta^n - \alpha^n \beta^m) \\
&= 2\alpha^n \beta^n (\alpha^{m-n} + \beta^{m-n}) \\
&= 2 b^n \bar{V}_{m-n}
\end{aligned}
\tag{5.13}
$$

We proved the identity $U_{2n} = U_n V_n$ for Lucas sequences. Let us check if it holds under the modification

$$\bar{U}_{2n} = \frac{\alpha^{2n} - \beta^{2n}}{\alpha^2 - \beta^2}$$

If n is odd, then

$$\bar{U}_{2n} = \left(\frac{\alpha^n - \beta^n}{\alpha - \beta}\right) \cdot \left(\frac{\alpha^n + \beta^n}{\alpha + \beta^n}\right)$$
$$= U_n V_n$$

Otherwise, n is even and we have

$$\bar{U}_{2n} = \left(\frac{\alpha^n - \beta^n}{\alpha^2 - \beta^2}\right) \cdot \left(\alpha^n + \beta^n\right)$$
$$= U_n V_n$$

So the identity holds for all n.

5.2 Divisibility of Lehmer Sequence

We can also use cyclotomic polynomials to describe (U) after modifying the definition of cyclotomic polynomials a little bit.

$$\Phi_n(\alpha, \beta) = \prod_{\substack{1 \le i \le n \\ \gcd(i,n)=1}} (\alpha - \zeta^i \beta)$$
$$= \beta^{\varphi(n)} \Phi_n\left(\frac{\alpha}{\beta}\right)$$

Durst [Dur61, Sect. 5] calls $\Phi_n(\alpha, \beta)$ the *Sylvester sequence* of (U). Then in the same fashion as $x^n - 1 = \prod_{d|n} \Phi_d(x)$ we have,

$$U_n = \prod_{\substack{d|n \\ d>1}} \Phi_d(\alpha, \beta)$$

This representation will be very important in our study. $\Phi_n(\alpha, \beta)$ follows the properties of the usual cyclotomic polynomial which we previously discussed in Sect. 1.4, Chap. 1. The same can be done for V_n if n is odd.

The following theorems are quite easy to establish.

Theorem 5.1 *The integers U_n and V_n are relatively prime to b.*

Carmichael [Car13] wrote the following:

$$(\alpha^n + \beta)^n = \alpha^n + \beta^n + \alpha\beta A$$
$$= V_n + bA$$
$$\frac{\alpha^n - \beta^n}{\alpha - \beta} = \alpha^{n-1} + \beta^{n-1} + \alpha\beta B$$
$$= V_{n-1} + bB$$

Now the proof is obvious. We will not consider α, β such that $\frac{\alpha}{\beta}$ is a root of unity. Otherwise, $U_n = $ if $\left(\frac{\alpha}{\beta}\right)^n = 1$ for some n. Conversely, if $U_n = 0$, then $\alpha^n = \beta^n$ and $V_n = 2\alpha^n$. Since V_n is relatively prime to $\alpha^n \beta^n$, $\alpha^n = \beta^n = 1$. The value of $|V_n|$ can be at most 2. Since $(\alpha - \beta)^2 = \Delta$, $\alpha - \beta \geq 1$. We can check the cases $U_n = 0$ and $V_0 = 0$ as well, however they turn out to be trivial cases like this. So, without loss of any essential generality, we can assume $U_n, V_n \neq 0$ and α, β are not roots of unity.

$$U_n = \frac{\alpha^n - \beta^n}{\alpha - \beta}$$
$$\leq (\alpha^n - \beta^n)$$
$$\leq |\alpha^n| + |\beta^n|$$
$$\leq 2$$

Theorem 5.2 $\gcd(U_n, V_n) \in \{1, 2\}$.

According to a footnote in Carmichael [Car13, Page 36], Lucas inaccurately stated in his original paper that $\gcd(U_n, V_n) = 1$.

Proof We can make use of the identity $(a + b)^2 - (a - b)^2 = 4ab$ and get

$$(\alpha^n + \beta^n)^2 - (\alpha^n - \beta^n)^2 = 4\alpha^n \beta^n$$
$$V_n^2 - \Delta U_n^2 = 4b^n$$

Let g_n be the greatest common divisor of U_n and V_n. Then $g_n^2 \mid U_n^2$ and $g_n^2 \mid V_n^2$, so we get $g_n^2 \mid 4b^n$. Evidently, we have $g_n^2 \mid 4$ or $g_n \mid 2$. However, we also have to show that $g_n = 2$ is possible. Carmichael uses the following cases to complete the proof. To show that both are possible, we can use $\alpha = 2, \beta = 1$ for 1 and $\alpha = 3, \beta = 1$ for 2. ∎

If we replace $\sqrt{c}$ by a, then $a \equiv 0 \pmod{m}$ implies that $c \equiv 0 \pmod{m^2}$. Similarly, $\sqrt{c} \equiv a \pmod{m}$ means that $c \equiv a^2 \pmod{m}$ so c is a quadratic residue of m. We will use Legendre symbol $\left(\frac{c}{m}\right)$ for quadratic residues (see Quadratic Residues in glossary for details).

Theorem 5.3 *U_n is divisible by 2 in the following cases:*

$$c = 4k, b = 2l + 1, n = 2h$$
$$c = 4k + 2, b = 2l + 1, n = 4h$$
$$c = 4k \pm 1, b = 2l + 1, n = 3h$$

V_n is divisible by 2 in the following cases:

$$c = 4k, b = 2l + 1$$
$$c = 4k + 2, b = 2l + 1, n = 2h$$
$$c = 4k \pm 1, b = 2l + 1, n = 3h$$

Proof Since U_n and V_n are prime to b^n, both are odd when b is even. So, in order for them to be even, b must be odd. Using the recurrences,

$$U_{n+2} = aU_{n+1} - bU_n$$
$$V_{n+2} = aV_{n+1} - bV_n$$

we have two cases: a is even or odd. If a is odd

$$U_{n+2} \equiv U_{n+1} + U_n \pmod 2$$
$$V_{n+2} \equiv V_{n+1} + V_n \pmod 2$$

In case a is even,

$$U_{n+2} \equiv U_n \pmod 2$$
$$V_{n+2} \equiv V_n \pmod 2$$

Since $U_1 = 1$, $U_2 = a$, if a is even then U_n is odd and even for odd and even n respectively. If a is odd, then we can easily see that U_n is even if $3 \mid n$, odd otherwise. Similarly, $V_1 = a$, $V_2 = a^2 - 2b = c - 2b$, so if c is even then V_n is even for all n. If c is odd then V_n is even if $3 \mid n$ and odd otherwise. We get the theorem using $c = a^2$ so a even implies that c is divisible by 4. ∎

We have already proved the following theorem. But we will show a proof inspired by Carmichael [Car13].

Theorem 5.4 (U) is a strong divisibility sequence and (V) is an oddly divisibility sequence.

Proof Assume that k is a divisor of n. Then we have the following.

$$\frac{U_n}{U_k} = \prod_{\substack{d \mid n \\ d \nmid k}} \Phi_d(\alpha, \beta)$$

$$U_m V_n - U_n V_m = 2b^n U_{m-n}$$

Since U_n and V_n are prime to b, any odd divisor of U_m and U_n would be a divisor of U_{m-n}. Then in a similar fashion as in Euclidean algorithm, we can use this to say as we did in Theorem 4.6, that odd divisor is a divisor of U_g as well where $g = \gcd(m, n)$. From the first of the two equations above, we see that U_g divides U_n since $\Phi_d(\alpha, \beta)$ is an integer polynomial in α, β. So, $\gcd(U_m, U_n)$ is $U_{\gcd(m,n)}$ if $\frac{U_m}{U_g}$ or $\frac{U_n}{U_g}$ is odd. Consider the pair $(\alpha^g, \beta^g) = (\bar{\alpha}, \bar{\beta})$ as roots of a quadratic equation.

$$\frac{U_n}{U_g} = \frac{\alpha^n - \beta^n}{\alpha^g - \beta^g}$$

$$= \frac{\bar{\alpha}^{\frac{n}{g}} - \bar{\beta}^{\frac{n}{g}}}{\bar{\alpha} - \bar{\beta}}$$

We again have $\gcd(\alpha^g \beta^g, \alpha^g + \beta^g) = 1$. Then, we can consider the corresponding sequences $(\bar{U}), (\bar{V})$ generated by $\bar{\alpha}, \bar{\beta}$. The same theorems above are applicable for these two modified sequences. If $\bar{\alpha}\bar{\beta}$ is even, both $\bar{U}_n$ and $\bar{V}_n$ are odd. If $\bar{\alpha}\bar{\beta}$ is odd and $\bar{\alpha} + \bar{\beta}$ is even, then one of $\bar{U}_{\frac{m}{g}}$ and $\bar{U}_{\frac{n}{g}}$ is odd since both $\frac{m}{g}$ and $\frac{n}{g}$ cannot be even. Again, if $\bar{\alpha}\bar{\beta}$ and $\bar{\alpha}\bar{\beta}$ are odd, one of $\bar{U}_{\frac{m}{g}}$ and $\bar{U}_{\frac{n}{g}}$ is odd since both cannot be divisible by 3. Thus, $\gcd(\bar{U}_{\frac{m}{g}}, \bar{U}_{\frac{n}{g}}) = 1$ and we have $\gcd(U_m, U_n) = U_g$. The proof of (V) is an oddly divisibility sequence follows immediately from the fact that $a + b$ divides $a^n + b^n$ for odd n.   ∎

Theorem 5.5 *Let m, n be positive integers and $g = \gcd(m, n)$ such that $\frac{m}{g}$ and $\frac{n}{g}$ are both odd. Then $\gcd(V_m, V_n) = V_{\gcd(m,n)}$.*

Proof We have the following identities.

$$U_{2m} = U_m V_m$$
$$U_{2n} = U_n V_n$$
$$U_{2\gcd(m,n)} = U_{\gcd(m,n)} V_{\gcd(m,n)}$$
$$U_{2g} = U_g V_g$$

So, V_g is a factor of U_{2g}. Since $\gcd(U_n, V_n) \mid 2$, we have that $\gcd(V_g, U_m) \mid 2$ and $\gcd(V_g, U_n \mid 2$. Therefore, we have either $\gcd(V_m, V_n) = V_g$ or $\gcd(V_m, V_n) = 2V_g$. We can show that the latter does not occur. To prove this, it is sufficient to show that either of $\frac{V_m}{V_g}$ or $\frac{V_n}{V_g}$ is odd. In a similar fashion as the proof above,

$$\frac{V_m}{V_g} = \frac{\alpha^m + \beta^m}{\alpha^g + \beta^g}$$

$$= \frac{\bar{V}_{\frac{m}{g}}}{\bar{V}_1}$$

Again, either $\frac{m}{g}$ or $\frac{n}{g}$ is not divisible by 3. If both $\bar{\alpha}\bar{\beta}$ and $\bar{\alpha} + \bar{\beta}$ are odd, then either $\bar{V}_{\frac{m}{g}}$ or $\bar{V}_{\frac{n}{g}}$ is odd. Next, consider that $\bar{\alpha}\bar{\beta}$ is odd and $\bar{\alpha} + \bar{\beta}$ is even. Then, $\alpha\beta$ is odd and $\alpha + \beta$ is even. Assume that $\alpha + \beta = 2^s l$ with l odd. Using $V_1 = a$ and $V_2 = a^2 - 2b = 2(2^{2s-1}l^2 - b)$, we can see that V_2 is of the form $2r$ for some odd r. Inductively, we can see that V_n is of the form $2^s l$ with l odd and $2r$ with r odd based on whether n is odd or even. Thus, both $\bar{V}_{\frac{n}{g}}$ and $\bar{V}_{\frac{m}{g}}$ are odd in this case, proving the theorem. ∎

In these proofs, we did not explicitly assume the Lehmer assumption. However, the proofs are almost similar so we will not go into subtle details in order to reduce tedious calculation like we did for establishing the identities a while ago. So here we will focus on the results in the original paper by Lehmer. Because if we just drop the extra a in U_n, then we are always dealing with integers. So our primary interest, in this case, is the integer coefficient of a.

Theorem 5.6 *Let n be a positive integer and p be a prime such that $p^a \parallel U_m$. If q is a positive integer not divisible by p, then for any positive integer l, U_{mqp^l} is divisible by p^{a+l}. Moreover, if $p^a \neq 2$, then $p^{a+l} \parallel U_{mqp^l}$.*

Lehmer called this result the *law of repetition.*

Proof We will make use of (5.10) to prove this. First, let us consider p odd. Every term in the expansion of U_{nk} in (5.10) has U_n at least. The term with least exponent of U_n in the expansion is for $i = \frac{k-1}{2}$. Setting k to p and using the fact that $pb^{\frac{n(p-1)}{2}}U_n$ has the least exponent of U_n in the expansion of U_{np}, we see that $p^{a+1} \parallel U_{mp}$. The reason is p^{a+i} divides all other terms where $i > 1$. By induction, we easily get that $p^{a+l} \parallel U_{np^l}$. If q is odd, set k to q in the equation and we get that U_{nq} has the same exponent of p in U_{np}. Otherwise, assume $q = 2^s r$ with r odd.

$$U_{nq} = U_{nr} V_{nr} V_{2nr} \cdots V_{2^{s-1}rn}$$

Since $\gcd(U_n, V_n)$ is either 1 or 2 whereas p is odd, $V_{2^i nr}$ has no odd divisor in common with U_{nq}. Similarly, as before $\nu_p(U_{nr}) = \nu_p(U_{nq})$ and $\nu_p(U_{nr}) = \nu_p(U_n)$.

For $p = 2$, we again see by (5.10) that $\nu_2(U_n q) = \nu_2(n)$. Also, V_{nq} is even if U_{nq} is even. Since $\gcd(U_n, V_n) \mid 2$, if U_{nq} is divisible by 2^a at most for $a > 1$, then $2^{a+1} \parallel U_{2nq}$. So, we have the theorem for all primes except when $p = 2, a = 1$. ∎

As a corollary, we have the following.

Theorem 5.7 *Let p be an odd prime and $\rho(p)$ be the rank of apparition of p in (U). If $\nu_p(U_\rho) = \alpha$, then the rank of apparition of $p^{\alpha+\lambda}$ is $\rho(p, \alpha + \lambda) = \rho(p) \cdot p^\lambda$. In other words, $p^{\alpha+\lambda} \parallel U_{\rho(p)p^\lambda}$.*

This is a generalization of an interesting property called *lifting the exponent.* We will prove later that if a and b are relatively prime such that the odd prime p divides $a - b$, then $\nu_p(a^n - b^n) = \nu_p(a - b) + \nu_p(n)$. This is known as *lifting the exponent lemma.* Clearly, the theorem we have just shown is a generalization of this exponent lifting property. We will get back to this in Chap. 6, Sect. 6.2. Durst [Dur61, Lemma 2] proves the following for $p^a = 2$ in Theorem 5.7.

Theorem 5.8 *If $2^a \parallel U_{2n}$ and $2^{a+1} \parallel U_{4n}$, then $2^{a+2} \parallel U_{8n}$.*

Recall that $n!_U$ is the product of the first n terms of the Lehmer sequence (U). We can calculate the exact power to which p divides $n!_U$ using (3.5).

$$\nu_p(n!_U) = \sum_{i \geq 1} \left\lfloor \frac{n}{\rho(p, i)} \right\rfloor$$

We have the following result.

Theorem 5.9 *Let p be a prime and $\rho = \rho(p, 1)$. Then we have $\rho(p, i) = \rho p^{i-1}$ and*

$$\nu_p(n!_U) = \sum_{i \geq 1} \left\lfloor \frac{n}{\rho p^{i-1}} \right\rfloor$$

If $p^\alpha \parallel U_\rho$ and $m = \lfloor n/\rho \rfloor$, then we also have

$$\nu_p(n!_U) = m\alpha + \sum_{i \geq 1} \left\lfloor \frac{m}{p^i} \right\rfloor$$

$$= m\alpha + \nu_p(m!)$$

Furthermore, if $m = b_k p^k + b_{k-1} p^{k-1} + \ldots + b_1 p + b_0$ is the base p representation of m, we have

$$\nu_p(n!_U) = m\alpha + \frac{m - (b_k + \ldots + b_0)}{p - 1}$$

For Fibonacci numbers, it is well known that $F_{p-\epsilon} \equiv 0 \pmod{p}$ where $\epsilon = \left(\dfrac{5}{p} \right)$ is the Legendre's symbol for odd prime p. This gives us a least upper bound on the rank of apparition of p in (F). Lucas proved a generalization in Lucas [Luc78a, Page 295–297]. Then Lehmer generalized this result even further [Leh30, Page 422–423]. For this generalization, let σ, ϵ be defined as $\epsilon = \left(\dfrac{\Delta}{p} \right), \sigma = \left(\dfrac{c}{p} \right)$ and p be an odd prime.

Theorem 5.10 *Let p be an odd prime. Then we have*

$$U_p \equiv \epsilon \pmod{p}$$
$$V_p \equiv \sigma a \pmod{p}$$

Proof Recall the identities

$$2\alpha = \sqrt{c} + \delta$$
$$2^p \alpha^p = \left(\sqrt{c} + \delta\right)^p$$
$$2\beta = \sqrt{c} - \delta$$
$$2^p \beta^p = \left(\sqrt{c} - \delta\right)^p$$

We can subtract and add to get U_p and V_p.

$$2^p \alpha^p = \sum_{i=0}^{p} \binom{p}{i} a^{p-i} \delta^i$$

$$2^p \beta^p = \sum_{i=0}^{p} \binom{p}{i} a^{p-i} (-\delta)^i$$

$$2^{p-1} U_p = \sum_{i=0}^{\frac{p-1}{2}} \binom{p}{2i+1} c^{\frac{p-2i-1}{2}} \Delta^i \tag{5.14}$$

$$2^{p-1} V_p = \sum_{i=0}^{\frac{p-1}{2}} \binom{p}{2i+1} a^{p-2i} \Delta^i \tag{5.15}$$

By Fermat's little theorem, $2^{p-1} \equiv 1 \pmod{p}$ and since $\binom{p}{i}$ is divisible by p for $0 < i < p$,

$$U_p \equiv \Delta^{\frac{p-1}{2}} \pmod{p}$$
$$\equiv \left(\frac{\Delta}{p}\right) \equiv \epsilon \pmod{p}, \text{ by Euler's criterion}$$
$$V_p \equiv c^{\frac{p}{2}} \pmod{p}$$
$$\equiv c^{\frac{1}{2}} c^{\frac{p-1}{2}} \pmod{p}$$
$$\equiv \sigma a \pmod{p}$$

∎

Theorem 5.11 *Let p be an odd prime. If bc is not divisible by p, then $U_{p-\sigma\epsilon} \equiv 0$ (mod p).*

Lehmer called this result the *law of apparition*.

Proof Set m to p, n to 1 in (5.6) and (5.7), respectively.

$$2b U_{p-1} = U_p V_1 - U_1 V_p$$
$$2b V_{p-1} = V_p V_1 - \Delta U_p U_1$$

Using $U_1 = 1$, $V_1 = a$, we have

$$2bU_{p-1} \equiv \epsilon a - a\sigma \pmod{p}$$
$$\equiv a(\sigma - \epsilon) \pmod{p}$$
$$\frac{2bU_{p-1}}{a} \equiv \sigma - \epsilon \pmod{p}$$
$$2bV_{p-1} \equiv c\sigma - \Delta\epsilon \pmod{p}$$

Again, setting m to p, n to 1 in (5.4) and (5.5) respectively,

$$2U_{p+1} = U_1 V_p + V_1 U_p$$
$$\equiv a\sigma + a\epsilon \pmod{p}$$
$$2V_{p+1} = V_p V_1 + \Delta U_1 U_p$$
$$\equiv a\sigma a + (c - 4b)\epsilon \pmod{p}$$
$$\equiv c(\sigma + \epsilon) - 4b\epsilon \pmod{p}$$

From (5.8) and (5.9),

$$U_{2p} \equiv \epsilon\sigma \pmod{p}$$
$$V_{2p} \equiv c\sigma^2 - 2b \pmod{p}$$

If $\sigma = \epsilon = 0$, then p divides both c and b. So this is not possible. Since $\left(\dfrac{a}{p}\right)$ can assume one of the values $\{-1, 0, 1\}$, there are 8 possible cases left based on the values of $\left(\dfrac{\Delta}{p}\right)$ and $\left(\dfrac{\sigma}{p}\right)$. We observe in the following table that the theorem follows.

ϵ	σ	U_p	V_{p-1}	$2V_{p+1}$	V_{2p}
1	1	1	2	$2c - 4b$	$c - 2b$
1	-1	1	$2 - \frac{c}{b}$	$-4b$	$c - 2b$
1	0	1	$2 - \frac{c}{2b}$	$c - 4b$	$-2b$
-1	1	-1	$-2 + \frac{c}{b}$	$4b$	$c - 2b$
-1	-1	-1	-2	$4c - 2b$	$c - 2b$
-1	0	-1	$-2 + \frac{c}{2b}$	$4b - c$	$-2b$
0	1	0	$\frac{c}{2b}$	c	$c - 2b$
0	-1	0	$-\frac{c}{2b}$	$-c$	$c - 2b$

■

Lucas [Luc78a, Page 297] proved the special case that $U_{p-\epsilon} \equiv 0 \pmod{p}$. If we take Δ such that Δ is a quadratic residue of p, then we have

$$U_{p-1} \equiv 0 \pmod{p}$$

$$\frac{\alpha^{p-1} - \beta^{p-1}}{\alpha - \beta} \equiv 0 \pmod{p}$$

$$\alpha^{p-1} \equiv \beta^{p-1} \pmod{p}$$

So this is a generalization of Fermat's little theorem. From Theorem 1.7, we see that for a prime p that does not divide bc, we have $\rho(p) \mid p - \sigma\epsilon$. For Fibonacci sequence, we know that $p \mid F_p$ only for $p = 5$, which makes sense because $\Delta = 5$ so $\left(\frac{5}{p}\right) = 0$. From this result, we get that $p \mid U_p$ if and only if $p \mid \Delta$ since then $\left(\frac{\Delta}{p}\right) = 0$. Moreover, if $p \mid b$, then $p \nmid U_n$ for all n. We can say the following in general.

Theorem 5.12 *Let p be an odd prime that does not divide bc. Then $\rho(p) \mid p - \sigma\epsilon$. Furthermore, we have the following:*

$$\rho(p) = \begin{cases} 2 & \text{if } p^2 \mid c \\ 2p & \text{if } p \mid c \text{ but } p^2 \nmid c \\ p & \text{if } p \mid \Delta \end{cases}$$

It is interesting to note that while the theorems above give us a least upper bound on the rank of the apparition; however, the problem is still equivalent to finding the least d such that $x^d \equiv 1 \pmod{p}$. Similarly, the problem of finding $\nu_p(U_\rho)$ is equivalent to finding $\nu_p(a^{p-1} - 1)$ which has not been completely solved either. Even so, we have the following result for the special case when $p \mid \Delta$.

Theorem 5.13 *If $p \mid \Delta$ for an odd prime $p > 3$, then $p^2 \nmid U_p$.*

Proof Taking modulo p^2 in (5.14),

$$2^{p-1}U_p \equiv pc^{\frac{p-1}{2}} + \binom{p}{3}c^{\frac{p-3}{2}}\Delta \pmod{p^2}$$

Using the fact that $p \mid \Delta$ and $p \mid \binom{p}{3}$ for odd prime $p > 3$.

$$2^{p-1}U_p \equiv pc^{\frac{p-1}{2}} \pmod{p^2}$$

Since $p \nmid c$, we have the proof for p^2 can not divide $pc^{\frac{p-1}{2}}$.   ∎

Let p be an odd prime and ρ be the rank of apparition of p in (U) such that ρ is odd. Then p cannot divide V_n. Because $U_{2n} = U_n V_n$ so $\rho \mid 2n$. Since ρ is odd, we get that $\rho \mid n$. Thus, p also divides U_n implying $p \mid \gcd(U_n, V_n)$. This is impossible since $\gcd(U_n, V_n) = 2$ and p is odd. We have the following result.

Theorem 5.14 *Let p be an odd prime such that the rank of apparition ρ of p in (U) is odd. Then p does not divide V_n for any n.*

We have the next result when the rank of the apparition is even.

Theorem 5.15 *Let p be an odd prime such that the rank of apparition $\rho = 2k$ of p in (U) is even. Then $V_{(2n+1)k}$ is divisible by p for every n and no other V_m is divisible by p.*

Proof Let m be a positive integer such that $p \mid V_m$. This can happen if and only if $\rho \mid m$. Using $U_{2m} = U_m V_m$, we have $\rho \mid 2m$. Using $\rho = 2k$, we have $k \mid m$. We show that p cannot divide V_{nk} where n is even. It is sufficient to show the case that V_{2k} is not divisible by p. For the sake of argument, assume that $p \mid V_{2k}$. Since $p \mid V_\rho$ and $p \mid U_\rho$, we again have $p \mid \gcd(U_\rho, V_\rho)$. A contradiction. For odd n, V_{nk} is divisible by V_k, so V_{nk} is divisible by p. The theorem follows.   ∎

Lucas also generalized Euler's totient function. For $n = p_1^{e_1} p_2^{e_2} \cdots p_r^{e_r}$, he defined the function $\psi(n)$ as

$$\psi(n) = \prod_{i=1}^{r} p_i^{e_i - 1} \left(p_i - \left(\frac{\Delta}{p_i} \right) \right)$$

Then for two roots α, β of the equation $x^2 - ux + v = 0$, Carmichael generalized it to

$$\varphi_{\alpha,\beta}(n) = \prod_{i=1}^{r} p_i^{e_i - 1} \left(p_i - \left(\frac{\alpha, \beta}{p_i} \right) \right), \quad \text{where}$$

$$\left(\frac{\alpha, \beta}{p} \right) \equiv (\alpha - \beta)^{p-1} \pmod{p}$$

Here, $\left(\frac{\alpha,\beta}{p} \right)$ is 0, 1 or -1 based on whether $(\alpha - \beta)^2$ is divisible by p or is a quadratic residue of p or a non-quadratic residue of p. Lehmer generalized this function even further.

$$\psi(n) = 2 \prod_{i=1}^{r} p_i^{e_i - 1} \left(p_i - \left(\frac{c\Delta}{p_i} \right) \right)$$

where $\left(\frac{c\Delta}{p} \right)$ is the usual Legendre symbol for an odd prime p. For $p = 2$,

$$\left(\frac{c\Delta}{2} \right) = \begin{cases} 0 & \text{if } c = 4k \\ -1 & \text{if } c = 2k + 1 \\ -2 & \text{if } c = 4k + 2 \end{cases}$$

The following result can be thought of as a generalization of Euler's totient theorem. It also gives an upper bound on the rank of apparition $\rho(n)$ for an arbitrary positive integer n.

Theorem 5.16 *Let n be a positive integer relatively prime to b. Then $U_{\psi(n)} \equiv 0$ (mod n), where*

$$\psi(n) = 2 \prod_{i=1}^{r} p_i^{e_i} p_i^{e_i-1} \left(p_i - \left(\frac{c\Delta}{p_i} \right) \right)$$

$$\left(\frac{c\Delta}{2} \right) = \begin{cases} 0 & \text{if } c = 4k \\ -1 & \text{if } c = 2k+1 \\ -2 & \text{if } c = 4k+2 \end{cases}$$

Evidently, $\rho(n) \mid \psi(n)$.

Proof If p is a prime factor of n and $p^e \parallel n$, then using the fact that

$$p - \left(\frac{\Delta}{p} \right) = p - \sigma\epsilon$$

we see that $p \mid U_{p-\sigma\epsilon}$. By the law of repetition, we have

$$U_{p^{e-1}(p-\sigma\epsilon)} \equiv 0 \pmod{p^e}$$

Since $\psi(n)$ is the product of all such $p^{e-1}(p - \sigma\epsilon)$, by the law of apparition, we have that $p^e \mid U_{\psi(n)}$ for all p. Thus, the least common multiple of all p^e, in other words, n divides $U_{\psi(n)}$. ∎

Note that we do not have to take the multiple of $p^{e-1}(p - \sigma\epsilon)$. If we take the least common multiple of them, we get an even lower upper bound than $\psi(n)$. This is actually a generalization of Carmichael's *universal exponent function*.

$$\lambda(n) = \text{lcm}_{1 \le i \le r} \left(p_i^{e_i-1} \left\{ p - \left(\frac{\sigma\epsilon}{p} \right) \right\} \right)$$

We have that $\rho(n) \mid \lambda(n)$ and $\lambda(n) \mid \psi(n)$. Moreover, we can actually take

$$\text{lcm} \left(2, \text{lcm}_{1 \le i \le r} \prod_{i=1}^{r} p_i^{e_i} p_i^{e_i-1} \left(p_i - \left(\frac{c\Delta}{p_i} \right) \right) \right)$$

instead of multiplying them since $\prod_{i=1}^{r} p_i^{e_i} p_i^{e_i-1} \left(p_i - \left(\frac{c\Delta}{p_i} \right) \right)$ may be even. So for a given positive integer we have a least index at which it appears as a factor in the sequence (U). The idea of primitive divisors arises if we consider the exact opposite. For a given index n, what would be the prime divisors of U_n? We can see that the interesting prime divisors would be the ones that did not appear as factors in the sequence before. Because otherwise we can generate those primes as divisors of previous terms. This makes us mostly interested in prime divisors, which are factors of U_n but not a factor of U_k for $k < n$. This is what we coined as *primitive prime divisor* of U_n. Lehmer [Leh30, Page 425] called a primitive divisor p of U_n *intrinsic* if $p \mid n$ holds as well. Otherwise, p is an *extrinsic* divisor. We will investigate the primitive divisors of U_n in Chap. 6. For now, we readily have the following using the law of apparition.

Theorem 5.17 *Let p be a primitive odd prime divisor of U_n. If p is extrinsic, then p is of the form $kn \pm 1$. If p is non-primitive, then $p \mid U_d$ for some divisor d of n such that $d < n$. Moreover, an odd extrinsic prime divisor p of V_n is of the form $2kn \pm 1$.*

5.3   Periodicity of Lehmer Sequence

We want to find the values of α and β such that (U) is periodic, that is, there is a ρ such that $U_{n+\rho} = U_n$ for all n. We will work under the conditions $\gcd(b, c) = 1$, $c > 0$, $\Delta \neq 0$ and $b \neq 0$. Our approach differs from the method Lehmer used in his paper [Leh30, Pages 425–426]. So, we will show both here.

We have two cases to consider. In the characteristic polynomial $f(x) = x^2 - ax + b$, we have $a = c^{\frac{1}{2}}$ either an integer or a an irrational. If a is integer by itself, then we have $f(x)$ is a polynomial of degree 2 irreducible in $\mathbb{Z}[x]$. Otherwise, we can get the minimal polynomial with integer coefficients that has α, β as roots by multiplying $f(x)$ with the conjugate.

$$
\begin{aligned}
g(x) &= (x^2 + b - ax)(x^2 + b + ax) \\
&= (x^2 + b)^2 - a^2 x^2 \\
&= x^4 + 2bx^2 + b^2 - cx^2 \\
&= x^4 - (c - 2b)x^2 + b^2
\end{aligned}
$$

Again, this is an irreducible polynomial in $\mathbb{Z}[x]$ and by Bell's theorem 2.6, (U) is periodic with period ρ if and only if α, β are primitive ρ-th roots of unity and $k = \varphi(\rho)$ where k is $\deg(f)$ if a is integer, $\deg(g)$ otherwise. Thus, (U) will be periodic for α, β if $\alpha + \beta = a$, $\alpha\beta = b$ and α, β are primitive ρ-th roots of unity. Since both α and β are roots of unity and $\alpha\beta$ is an integer, we must have that α and β are reciprocal of each other. Let ζ be a primitive ρ-th root of unity. Then the roots of unity are ζ^i for $0 \leq i < \rho$. The primitive roots are of the form ζ^i where $\gcd(i, \rho) = 1$. If $\alpha = \zeta^s$ for $\gcd(s, \rho) = 1$, then $\beta = \rho - s$. Since $\alpha \neq \beta$, we can assume without loss of generality $s < \rho - s$. Also, both α and β are roots of unity, so by taking modulus we see that b must have absolute value 1.

Lehmer reaches the same conclusion that $|b| = 1$. However, his approach was to consider the congruence $U_{p-\sigma\epsilon} \equiv 0 \pmod{p}$. Since (U) is periodic, this congruence holds for all prime. But due to the periodicity, the absolute value of U_n cannot be arbitrarily large. Therefore, for a prime p that is larger than this value of U_n, $U_{p-\epsilon\sigma}$ must be essentially 0. Then we get $\alpha^r - \beta^r = 0$ for some r where $\alpha \neq \beta$. Using the fact that $(\alpha + \beta)^2$ and $\alpha\beta$ are relatively prime, we reach that α and β are roots of unity.

We get two cases. Lehmer handles these cases as stated below.

1. $b = 1$. Lehmer says that in this case $\alpha = e^{\frac{2i\pi k}{m}}$ and $\beta = e^{\frac{-2i\pi k}{m}}$. However, from Lehmer's conclusion, we do not necessarily require α and β to be primitive roots of unity as long as they are reciprocal of each other.

$$c = (\alpha + \beta)^2$$
$$= \left(e^{\frac{2i\pi k}{m}} + e^{\frac{-2i\pi k}{m}}\right)^2$$
$$= \left(\cos\frac{2\pi k}{m} + i\sin\frac{2\pi k}{m} + \cos\frac{2\pi k}{m} - i\sin\frac{2i\pi k}{m}\right)^2$$
$$= 4\cos^2\frac{2\pi k}{m}$$

In order for c to be integer, $\cos^2\frac{2\pi k}{m}$ has to be either an integer or one of the values in $\{\frac{1}{2}, \frac{1}{4}, \frac{3}{4}\}$. Since $\alpha \neq \beta$, we cannot have $k = 0$, so $c \in \{1, 2, 3\}$. $c \neq 0$ since $\gcd(c, b) = 1$.

2. $b = -1$. In this case, $\alpha = e^{\frac{2i\pi k}{m}}$ and $\beta = -e^{\frac{-2i\pi k}{m}}$. In a similar fashion as before, $c = -4\sin^2\frac{2k\pi}{m}$. We get $c \in \{-1, -2, -3\}$.

Since there are 6 cases for c, we get a total of 12 pairs of (α, β) since $a = c^{\frac{1}{2}}$ can assume two values. Lehmer shows the following table describing all pairs along with the period (U) assume, however, he does not show the process how he derived the values.

Here is Table 5.1 with the values of α, β which produce periodic (U) along with their period ρ.

Since Lehmer does not show how he calculates the values ρ, we will continue where we left off and deduce the values of α, β, ρ. We deduced that α, β are primitive ρ-th roots of unity and $|b| = 1$. From Bell's theorem, we must have $k = \varphi(\rho)$. From our discussion, $k \in \{2, 4\}$. If $k = 2$, then the possible values of ρ is in the set $\{3, 4, 6\}$. If $k = 4$, then $\rho \in \{5, 8, 10, 12\}$. In order to further restrict the values of ρ, we let $\zeta = e^{\frac{2i\pi}{\rho}}$.

Table 5.1 Values of α, β and ρ for periodic (U)

b	a	Δ	α	β	ρ
1	1	-3	$e^{\frac{2i\pi}{6}}$	$e^{\frac{-2i\pi}{6}}$	6
1	-1	-3	$e^{\frac{2i\pi}{3}}$	$e^{\frac{-2i\pi}{3}}$	3
-1	i	3	$e^{\frac{2i\pi}{12}}$	$e^{\frac{-2i\pi}{12}}$	6
-1	$-i$	3	$e^{\frac{2i\pi}{12}}$	$e^{\frac{-2i\pi}{12}}$	12
1	$\sqrt{2}$	-2	$e^{\frac{2i\pi}{8}}$	$e^{\frac{-2i\pi}{8}}$	8
1	$-\sqrt{2}$	-2	$-e^{\frac{-2i\pi}{8}}$	$-e^{\frac{2i\pi}{8}}$	8
-1	$\sqrt{2}i$	2	$e^{\frac{2i\pi}{8}}$	$-e^{\frac{-2i\pi}{8}}$	8
-1	$-\sqrt{2}i$	2	$e^{\frac{-2i\pi}{8}}$	$e^{\frac{2i\pi}{8}}$	8
1	$\sqrt{3}$	-1	$e^{\frac{2i\pi}{12}}$	$e^{\frac{-2i\pi}{12}}$	12
1	$-\sqrt{3}$	-1	$-e^{\frac{-2i\pi}{12}}$	$-e^{\frac{2i\pi}{12}}$	12
-1	$\sqrt{3}i$	1	$-e^{\frac{2i\pi}{6}}$	$e^{\frac{-2i\pi}{6}}$	6
-1	$-\sqrt{3}i$	1	$e^{\frac{-2i\pi}{6}}$	$-e^{\frac{2i\pi}{6}}$	6

$$c = (\alpha + \beta)^2$$
$$= \left(e^{\frac{2i\pi s}{\rho}} + e^{\frac{2i\pi(\rho-s)}{\rho}}\right)^2$$
$$= \left(e^{\frac{2i\pi s}{\rho}} + e^{\frac{-2i\pi s}{\rho}}\right)^2$$
$$= 4\cos^2\frac{2\pi s}{\rho}$$

This is the same as Lehmer's result but this time we know the meaning of s and ρ. We can derive the values of s and ρ proceeding as before.

5.4 Extending Wilson's Theorem

Lehmer [Leh30, Page 426] calls a (U) *degenerate* if $\Delta = 0$. Then $c = 4b$, $a = 2\sqrt{b}$ and $\alpha = \beta = \frac{a}{2} = b$. In this case, $\frac{\alpha^n - \beta^n}{\alpha - \beta}$ is undefined, so we have

$$U_n = \lim_{\alpha \to \beta} \frac{\alpha^n - \beta^n}{\alpha - \beta}$$
$$= n\beta^{n-1}$$
$$V_n = 2\alpha^n$$

Since $\gcd(b, c) = 1$, we must have $c = \pm 4$, $b = \pm 1$. We get the following degenerate cases.

i. $a = 2, b = 1, U_n = n, V_n = 2$.
ii. $a = -2, b = 1, U_n = n(-1)^{n-1}, V_n = 2(-i)^n$.
iii. $a = 2i, b = -1, U_n = ni^{n-1}, V_n = 2i^n$.
iv. $a = -2i, b = -1, U_n = n(-1)^{n-1}, V_n = 2(-i)^n$.

(U) is *degenerate modulo m* if $bc\Delta \equiv 0 \pmod{m}$. Wilson's theorem states that for any prime p, $(p-1)! \equiv -1 \pmod{p}$. For a proof, one can see Billal and Hossein [BH18, Theorem 2.6.1, Page 87]. Gauss [Gau01, Article 78] also states this result. For degenerate (U) modulo m, Wilson's theorem can be extended as follows.

Theorem 5.18 (Extension of Wilson's theorem) *Let p be an odd prime divisor of Δ. Then for $(\bar{U})$ defined in (5.1),*

$$\prod_{i=1}^{p-1} \bar{U}_i \equiv -\left(\frac{2}{p}\right)\sigma^{\frac{p-3}{2}} \pmod{p}$$

Proof If n is even, then $\bar{U}_n = \frac{U_n}{a}$, otherwise $\bar{U}_n = U_n$. Let m be an odd divisor of Δ. Taking modulo m in (5.14),

$$2^{i-1}U_i \equiv ic^{\frac{i-1}{2}} \pmod{m}$$

Setting $m \to p$, we get

$$\prod_{i=1}^{p-1} \bar{U}_i \equiv \left(\prod_{i=1}^{\frac{p-1}{2}} \frac{U_{2i}}{\sqrt{c}}\right)\cdot\left(\prod_{i=1}^{\frac{p-3}{2}} U_{2i}\right) \pmod{p}$$

$$\equiv \left(\prod_{i=1}^{\frac{p-1}{2}} \frac{2i}{2^{1-2i}}c^{i-1}\right)\cdot\left(\prod_{i=1}^{\frac{p-3}{2}} (2i+1)2^{-2i}c^i\right) \pmod{p}$$

Collecting terms, we get

$$\prod_{i=1}^{p-1} \bar{U}_i \equiv (p-1)!2^{-\frac{(p-1)(p-2)}{2}}c^{\frac{(p-1)(p-3)}{4}} \pmod{p}$$

$$\equiv -\left(\frac{2}{p}\right)\left(\frac{c}{p}\right)^{\frac{p-3}{2}} \pmod{p}$$

∎

Note that, using $\left(\frac{c}{p}\right)^2 = 1$, we can also replace $\sigma^{\frac{p-3}{2}}$ by $\sigma^{\frac{p+1}{2}}$. This can be generalized based on whether m is odd or even. For this generalization, we will need the following results. A proof of these can be found in Billal and Hossein [BH18, Page 126–127, Theorems 2.12.22, 2.12.23].

Theorem 5.19 *Let $n > 1$ be a positive integer. Then there is a primitive root modulo n if and only if $n \in \{2, 4, p^{\alpha}, 2p^{\alpha}\}$ for some odd prime p and a positive integer α.*

Theorem 5.20 *Let n be a positive integer and $\mathbb{U}_n$ be the set $\{i \in \mathbb{N} : \gcd(i, n) = 1, 1 \leq i \leq n\}$. If the elements of $\mathbb{U}_n$ are $g_1, g_2, \ldots, g_{\varphi(n)}$, then*

$$g_1 g_2 \cdots g_{\varphi(n)} \equiv \begin{cases} -1 & (\mathrm{mod}\ n) & \textit{if there exists a primitive root modulo } n \\ 1 & (\mathrm{mod}\ n) & \textit{otherwise} \end{cases}$$

Lehmer denoted $a^{\frac{\varphi(n)}{2}}$ by $\left\{\dfrac{a}{m}\right\}$ in the next theorem.

Theorem 5.21 *Let m be an odd divisor of Δ. Then*

$$\prod_{\substack{1 \leq i \leq m \\ \gcd(i,m)=1}} \bar{U}_i \equiv \gamma \left\{\frac{2}{m}\right\} \left\{\frac{c}{m}\right\}^{\frac{m+1}{2}} \quad (\mathrm{mod}\ m), \textit{ where}$$

$$\gamma = \begin{cases} -1 & \textit{if } m = p^r \textit{ for an odd prime } p \\ 1 & \textit{otherwise} \end{cases}$$

Proof Let E and O, respectively, be the set of even and odd positive integers that are less than or equal to m and relatively prime to m.

$$\prod_{\substack{1 \leq i \leq m \\ \gcd(i,m)=1}} \bar{U}_i = \prod_{i \in E} U_i \prod_{j \in O} \bar{U}_j$$

$$\equiv \prod_{i \in E} i 2^{1-i} c^{\frac{i-1}{2}} \prod_{j \in O} j 2^{1-j} c^{\frac{j-2}{2}} \quad (\mathrm{mod}\ m)$$

$$\equiv 2^{\sum_{i \in E}(1-i)+\sum_{j \in O}(1-j)} c^{\frac{\sum_{i \in E}(i-1)+\sum_{j \in O}(j-2)}{2}} \prod_{\substack{1 \leq i \leq m \\ \gcd(i,m)=1}} i \quad (\mathrm{mod}\ m)$$

If a is relatively prime to n, then so is $n - a$. Hence $\varphi(n)$ is even for $n > 2$, so in the sum $\sum_{i \in E} i + \sum_{j \in O} j$, we can pair up and get $\frac{\varphi(n)}{2}$ pairs of n. Using this and Theorem 5.20, we have

$$\sum_{i \in E} i + \sum_{j \in O} j = \frac{n\varphi(n)}{2}$$

$$\prod_{\substack{1 \leq i \leq m \\ \gcd(i,m)=1}} i \equiv \gamma \quad (\mathrm{mod}\ m)$$

where γ is -1 if m has a primitive root else 1.

$$\prod_{\substack{1\le i\le m \\ \gcd(i,m)=1}} \bar{U}_i \equiv \gamma 2^{-\frac{m\varphi(m)}{2}+\varphi(m)} c^{\frac{\varphi(m)(m-3)}{4}} \pmod{m}$$

$$\equiv \gamma 2^{-\frac{m\varphi(m)}{2}} 2^{\varphi(m)} c^{\frac{\varphi(m)(m-3)}{4}} c^{\varphi(m)} \pmod{m}$$

$$\equiv \gamma 2^{-\frac{m\varphi(m)}{2}} c^{\frac{\varphi(m)(m+1)}{4}} \pmod{m}$$

$$\equiv \gamma \left\{\frac{2}{m}\right\} \left\{\frac{c}{m}\right\}^{\frac{m+1}{2}} \pmod{m}$$

Here $\gamma = -1$ if $n = p^\alpha$ for an odd prime p and a positive integer α, since m can not be even. ∎

Theorem 5.22 *Let m be an even divisor of Δ.*

$$\prod_{\substack{1\le i\le m \\ \gcd(i,m)=1}} \bar{U}_i \equiv \gamma' \left\{\frac{b}{m}\right\}^{\frac{m-2}{2}} \pmod{m}$$

where $\gamma' = -1$ if $m \in \{2, 4, 2p^\alpha\}$ for an odd prime p, otherwise $\gamma' = 1$.

Proof Since m is even, all the positive integers relatively prime to m are odd.

$$\prod_{\substack{1\le i\le m \\ \gcd(i,m)=1}} \bar{U}_i \equiv \prod_{\substack{1\le i\le m \\ \gcd(i,m)=1}} ic^{\frac{i-1}{2}} \pmod{m}$$

$$\equiv \gamma' c^{\frac{\sum_{\substack{1\le i\le n \\ \gcd(i,m)=1}}(i-1)}{2}} \pmod{m}$$

$$\equiv \gamma' c^{\frac{\frac{m\varphi(m)}{2}-\varphi(m)}{2}} \pmod{m}$$

$$\equiv \gamma' c^{\frac{(m-2)\varphi(m)}{4}} \pmod{m}$$

$$\equiv \gamma' \left\{\frac{c}{m}\right\}^{\frac{m-2}{2}} \pmod{m}$$

∎

5.5 LCM Sequence of Lehmer Sequences

We will now consider the lcm sequence of (U). We have already established that (U) can be written as

$$U_n = \prod_{d|n} \Phi_d(\alpha, \beta)$$

where

$$\Phi_n(\alpha, \beta) = \prod_{\substack{1 \le j \le n \\ \gcd(j,n)=1}} \left(\alpha - e^{\frac{2i\pi j}{n}} \beta\right)$$

These factors $\Phi_d(\alpha, \beta)$ are the terms of the lcm sequence of (U). However, a special meaning of $\Phi_n(\alpha, \beta)$ for U_n is that it is the product of the primitive prime divisors of U_n. Because other divisors are already found in $\Phi_d(\alpha, \beta)$ for $d < n$. From Theorem 3.11,

$$\Phi_n(\alpha, \beta) = \frac{U_n}{\mathrm{lcm}(U_{\frac{n}{p_1}}, U_{\frac{n}{p_2}}, \ldots, U_{\frac{n}{p_r}})}$$

where $p_1, p_2, \ldots, p_r$ are distinct prime factors of n. Consider a prime p. We want to investigate the rank of apparition of p in (Φ), where $\Phi_n = \Phi_n(\alpha, \beta)$. If ρ is the rank of apparition of p in (Φ), then $p \mid \Phi_\rho$. So, p must divide U_ρ as well, hence ρ is the same as the rank of apparition of p in (U).

Theorem 5.23 *For a prime p, the rank of apparition of p in (Φ) is the same as the rank of apparition in (U).*

We want to characterize all the indexes s such that $p \mid \Phi_s$. For discarding triviality, we will assume $s \nmid 6$ and $s \neq 4$. If $p \mid \Phi_s$, then we have $\rho \mid s$. Since $p \mid \Phi_\rho$ and $p \mid \Phi_s$, we have $\gcd(\Phi_s, \Phi_\rho) > 1$. By Theorem 1.48, we can see $\frac{s}{\rho}$ must be a power of p. So, we have the next theorem

Theorem 5.24 *Let p be an odd prime and the rank of apparition of p in (Φ) be ρ. Then for a positive integer s, $p \mid \Phi_s$ if and only if $\frac{s}{\rho}$ is a power of p.*

Lehmer [Leh30, Page 431] proves this result as well. However, there is a concern regarding this proof. Since $\rho \mid s$, letting $s = \rho k$ and $k = p^\lambda l$ where $\lambda \ge 0$ and $\gcd(l, p) = 1$, we have

$$\begin{aligned}
U_s = U_{k\rho} \\
= \prod_{d|k\rho} \Phi_d(\lambda, \beta) \\
= \prod_{d|\rho} \Phi_d(\alpha, \beta) \prod_{\substack{d \nmid \rho \\ d|k\rho}} \Phi_d(\alpha, \beta) \\
= U_\rho \prod_{\substack{d \nmid \rho \\ d|k\rho}} \Phi_d(\alpha, \beta)
\end{aligned}$$

Thus, $\Phi_{k\rho}$ divides $\frac{U_s}{U_\rho}$. Since $p \mid U_\rho$ and $p \nmid l$, by Theorem 5.7,

$$\nu_p(U_{k\rho}) = \nu_p(U_{p^\lambda l\rho})$$
$$= \nu_p(U_\rho) + \lambda$$
$$\nu_p\left(\frac{U_s}{U_\rho}\right) = \nu_p(U_s) - \nu_p(U_\rho)$$
$$= \nu_p(U_\rho) + \lambda - \nu_p(U_\rho)$$
$$= \lambda$$

If $\lambda = 0$, then p does not divide $\frac{U_s}{U_\rho}$, so p cannot divide $\Phi_{k\rho}$. This is impossible, so $\lambda > 0$ and p must divide k. Next, if $l > 1$, we have

$$U_{\rho p^\lambda l} = \prod_{d \mid \rho p^\lambda l} \Phi_d(\alpha, \beta)$$
$$= \prod_{d \mid \rho p^\lambda} \Phi_d(\alpha, \beta) \prod_{\substack{d \mid \rho p^\lambda l \\ d \nmid \rho p^\lambda}} \Phi_d(\alpha, \beta)$$
$$= U_{\rho p^\lambda} \prod_{\substack{d \mid \rho p^\lambda l \\ d \nmid \rho p^\lambda}} \Phi_d(\alpha, \beta)$$
$$\frac{U_{\rho p^\lambda l}}{U_{\rho p^\lambda}} = \prod_{\substack{d \mid \rho p^\lambda l \\ d \nmid \rho p^\lambda}} \Phi_d(\alpha, \beta)$$

Lehmer states that $\Phi_{\rho p^\lambda}(\alpha, \beta)$ divides $\frac{U_{\rho p^\lambda l}}{U_{\rho p^\alpha}}$. The notation Lehmer used $(G_{\omega p^\lambda})$ is different but the meaning is same. His exact statement is quoted below.

First, let p be odd and ω be its rank of apparition. If p divides G_r, then r is of the form $k\omega$, since otherwise $G_r \neq 0 \pmod{p}$. Furthermore, k is divisible by p, for G_{kw} is a divisor of U_{kw}/U_w which by Theorem 1.6 does not contain the factor p unless k is divisible by p. Let $k = p^\lambda k'$, $k' > 1$ and prime to p. Then $G_{\omega p^\lambda}$ is a divisor of $U_{\omega p^\lambda k'}/U_{\omega p^\lambda}$ which by Theorem 1.6 does not contain the factor p, since k' is prime to p. We therefore consider the case $r = \omega p^\lambda$.

The claim that $\Phi_{\rho p^\lambda}$ is a divisor of $U_{\rho p^\lambda l}/U_{\rho p^\lambda}$ is not necessarily true. Because $\Phi_{\rho p^\lambda}$ is divisible by p whereas $U_{\rho p^\lambda l}/U_{\rho p^\lambda}$ is not by Theorem 5.7. He does not provide additional argument for why he thinks that $\Phi_{\rho p^\lambda}(\alpha, \beta)$ has to divide $\frac{U_{\rho p^\lambda l}}{U_{\rho p^\lambda}}$.

Now, we have that p divides Φ_s if and only if $s = \rho p^\lambda$. Suppose that $\rho \neq p$ and $\rho \neq 2p$, $\lambda > 1$.

$$\frac{U_{\rho p^\lambda}}{U_{\rho p^{\lambda-1}}} = \prod_{\substack{d\mid\rho p^\lambda \\ d\nmid\rho p^{\lambda-1}}} \Phi_d(\alpha, \beta)$$

$$= \Phi_{\rho p^\lambda}(\alpha, \beta) \prod_{\substack{d\mid\rho \\ d<\rho}} \Phi_{dp^\lambda}(\alpha, \beta)$$

By Theorem 5.7,

$$\nu_p\left(\frac{U_{\rho p^\lambda}}{U_{\rho p^{\lambda-1}}}\right) = 1$$

Since $p \mid \Phi_{\rho p^\lambda}$ and p^2 does not divide the left side of this equation, $\Phi_{\rho p^\lambda}$ is divisible by p but not by p^2. If $\rho = p$, then by Theorem 3.11,

$$\Phi_{\rho p^\lambda}(\alpha, \beta) = \frac{U_{\rho p^\lambda}}{U_{\rho p^{\lambda-1}}}$$

Again, if $\rho \in \{2, 2p\}$,

$$U_{2p^\lambda} = \prod_{d\mid 2p^\lambda} \Phi_d(\alpha, \beta)$$

$$= \Phi_{2p^\lambda} \prod_{d\mid p^\lambda} \Phi_d(\alpha, \beta) \prod_{\substack{d\mid 2p^\lambda \\ d\nmid p^\lambda \\ d<2p^\lambda}} \Phi_d(\alpha, \beta)$$

$$= \Phi_{2p^\lambda} U_{p^\alpha} \prod_{d\mid p^{\lambda-1}} \Phi_{2d}(\alpha, \beta)$$

$$= \Phi_{2p^\lambda} U_{p^\lambda} \frac{\prod_{d\mid 2p^{\lambda-1}} \Phi_d(\alpha,\beta)}{\prod_{d\mid p^{\lambda-1}} \Phi_d(\alpha,\beta)}$$

$$= \Phi_{2p^\lambda} U_{p^\lambda} \frac{U_{2p^{\lambda-1}}}{U_{p^{\lambda-1}}}$$

$$\Phi_{2p^\lambda} \frac{U_{p^\lambda}}{U_{p^{\lambda-1}}} = \frac{U_{2p^\lambda}}{U_{2p^{\lambda-1}}}$$

Since ρ does not divide p^i, $p \nmid \frac{U_{p^\lambda}}{U_{p^{\lambda-1}}}$. By Theorem 5.7, $p \parallel \frac{U_{2p^\lambda}}{U_{2p^{\lambda-1}}}$ so we have $p \parallel \Phi_{2p^\lambda}(\alpha, \beta)$.

If $p = 2$, then p divides one of U_2, U_3, or U_4. In a similar manner as above, we have that $\Phi_{2^\lambda}(\alpha, \beta)$ or $\Phi_{3\cdot2^\lambda}(\alpha, \beta)$ can be even and other $\Phi_n(\alpha, \beta)$ will be odd. Combining these results, we have the next theorem.

Theorem 5.25 *Let n be a positive integer such that $n \nmid 6$, $n \nmid 4$ and ρ be the rank of apparition of a prime p in (U). Then $\Phi_n(\alpha, \beta)$ is divisible by p if and only if $n = \rho p^\lambda$*

and $p^2 \nmid \Phi_n(\alpha, \beta)$ if $\lambda > 0$. If $\lambda = 0$, then $\Phi_n(\alpha, \beta)$ may possibly be divisible by p^2 except when $\rho = p$ or $\rho = 2p$ for a prime $p > 3$. $\Phi_n(\alpha, \beta)$ is even if and only if $n \in \{2^\lambda, 3 \cdot 2^\lambda\}$ and $2 \parallel \Phi_n(\alpha, \beta)$.

5.6 Subsequence of Lehmer Sequences

Consider a subsequence $(U)^{(r)}$ of (U) defined as $U_n^{(r)} = \frac{U_{nr}}{U_r}$. $(U)^{(r)}$ can be rearranged in terms of α and β as follows.

$$U_n^{(r)} = \frac{\alpha^{nr} - \beta^{nr}}{\alpha^r - \beta^r}$$

So $(U)^{(r)}$ is the Lehmer sequence for α^r, β^r, where α, β are the roots. Since $\alpha^r + \beta^r = V_r$ and $\alpha^r \beta^r = b^r$, α^r, β^r are the solutions of the equation $x^2 - V_r x + b^r = 0$, which is the characteristic equation for $(U)^{(r)}$. Thus, $U_{n+2}^{(r)} = V_r U_{n+1}^{(r)} - b^r U_n^{(r)}$ due to the characteristic equation of $(U)^{(r)}$. So, $(U)^{(r)}$ is also a second-order linear divisibility sequence. To compare it with Lehmer sequence, $c' = V_r^2, a' = V_r, b' = b^r$ and $\alpha' = \alpha^r, \beta' = \beta^r, \Delta' = \Delta U_r^2, \delta' = \delta U_r$.

$(U)^{(r)}$ is a divisibility sequence for if $m \mid n$,

$$\frac{U_n^{(r)}}{U_m^{(r)}} = \frac{\alpha^{nr} - \beta^{nr}}{\alpha^{mr} - \beta^{mr}}$$

is an integer. More specifically, it is a strong divisibility sequence.

$$\begin{aligned}
\gcd(U_n^{(r)}, U_m^{(r)}) &= \gcd\left(\frac{\alpha^{nr} - \beta^{nr}}{\alpha^r - \beta^r}, \frac{\alpha^{mr} - \beta^{mr}}{\alpha^r - \beta^r}\right) \\
&= \frac{\gcd(\alpha^{nr} - \beta^{nr}, \alpha^{mr} - \beta^{mr})}{\alpha^r - \beta^r} \\
&= \frac{\alpha^{\gcd(nr,mr)} - \beta^{\gcd(nr,mr)}}{\alpha^r - \beta^r} \\
&= \frac{\alpha^{r \gcd(m,n)} - \beta^{r \gcd(m,n)}}{\alpha^r - \beta^r} \\
&= U_{\gcd(m,n)}^{(r)}
\end{aligned}$$

The sequence $(U)^{(r)}$ remains unchanged for $r = 1$. An interesting case is $r = 2k$ since V_{2k} is an integer. In this case,

$$U_n^{(2k)} = V_{2k} U_{n-1}^{(2k)} - b^r U_{n-2}^{(2k)}$$

So, $(U)^{(2k)}$ is a Lucas sequence of the first kind.

Theorem 5.26 *Let p be an odd prime and ρ be the rank of apparition of p in (U). Then the rank of apparition of p in $(U)^r$ is*

$$\rho^{(r)} = \begin{cases} p & \text{if } \rho \mid r \\ \dfrac{\mathrm{lcm}(r, \rho)}{r} & \text{otherwise} \end{cases}$$

Proof U_n is divisible by p if and only if $\rho \mid n$. Now, considering $(U)^{(r)}$ as taking every r-th term of (U), in order for p to divide U_n^r, we need both $p \mid U_n$ and $r \mid n$. Thus, $\rho \mid n$ and $r \mid n$ and the least possible index for which this is true is $= \mathrm{lcm}(r, \rho)$. However, since only every r-th term is taken, the number of terms is reduced by $1/r$-th as well. So, the rank of apparition of p in $(U)^{(r)}$ is $\frac{\mathrm{lcm}(\rho,r)}{r}$. In the case $\rho \mid r$, we have that $\Delta^{(r)} = \Delta U_r^2$ is divisible by p. So, $\left(\dfrac{\Delta}{p}\right) = 0$ and by Theorem 5.12, $\rho = p$. ∎

Theorem 5.27 *If $p \nmid U_{2r}$, then $\sigma\epsilon$ remains unchanged for $(U)^{(r)}$.*

Proof Letting $\tau = \left(\dfrac{b}{p}\right)$, we have

$$\sigma^{(r)} = \left(\frac{V_r^2}{p}\right)$$

$$\epsilon^{(r)} = \left(\frac{\Delta U_r^2}{p}\right)$$

$$\tau^{(r)} = \left(\frac{b^r}{p}\right)$$

Since $p \nmid U_{2r}$ and $U_{2r} = U_r V_r$, we have $p \nmid U_r$ and $p \nmid V_r$.

$$\sigma^{(r)} = \begin{cases} \sigma & \text{if } r \text{ is odd} \\ 1 & \text{otherwise} \end{cases}$$

$$\epsilon^{(r)} = \begin{cases} \epsilon & \text{if } r \text{ is odd} \\ \sigma\epsilon & \text{otherwise} \end{cases}$$

$$\tau^{(r)} = \begin{cases} \tau & \text{if } r \text{ is odd} \\ 1 & \text{otherwise} \end{cases}$$

Thus, the product $\sigma\epsilon$ does not change. ∎

References

[BH18]   M. Billal, A. Hossein, *Topics in Number Theory: An Olympiad-Oriented Approach*, 1st edn. (Amazon, 2018)

[Car13]   R.D. Carmichael, On the numerical factors of the arithmetic forms $\alpha^n \pm \beta^n$. Ann. Mat. **15**(1/4), 30–48 (1913). ISSN: 0003486X, http://www.jstor.org/stable/1967797

[Dur61]   L.K. Durst, Exceptional real Lucas sequences. Pacific J. Math. **11**(2), 489–494 (1961), https://projecteuclid.org:443/euclid.pjm/1103037329

[Gau01]   C.F. Gauss, Disquisitiones arithmeticae (1801)

[Leh30]   D.H. Lehmer, An extended theory of Lucas' functions. Ann. Mat. **31**(3), 419–448 (1930). ISSN: 0003486X, http://www.jstor.org/stable/1968235

[Luc78a] E. Lucas, Théorie des Fonctions Numériques Simplement Périodiques. Am. J. Math. **1**(2), 184–196 (1878). ISSN: 00029327, 10806377, https://www.jstor.org/stable/2369308

[Luc78b] E. Lucas, Théorie des Fonctions Numériques Simplement Périodiques. Am. J. Math. **1**(3), 197–240 (1878). ISSN: 00029327, 10806377, http://www.jstor.org/stable/2369311

[Luc78c] E. Lucas, Théorie des Fonctions Numériques Simplement Périodiques. Am. J. Math. **1**(4), 289–321 (1878). ISSN: 00029327, 10806377, http://www.jstor.org/stable/2369373

Chapter 6
On Primitive Divisors

This chapter is dedicated to investigating special divisors of the sequences we have
discussed so far. More specifically, we are mostly interested in primitive divisors of
such sequences. We will first see a chronological history of this literature and then
go through each one of them sequentially. We will also show some results regarding
the size of primitive divisors and the number of distinct primitive divisors. In case
you are confused about why we are obsessed with primitive divisors, this is a natural
question that arises when we consider the rank of the apparition of a prime in a
sequence. If ρ is the rank of the apparition of p in (U), then $p \mid U_\rho$, but what if we
are given a certain index n? What can be said about the prime divisors of U_n? Does
it contain any factor that is *exclusive* to U_n? If not, then all the prime factors of U_n
have appeared before in the sequence, so they are not of any particular interest. In
the other case, we have a prime p that has appeared in the sequence for the first
time. The other question that follows is whether it is necessary for every term of the
sequence to contain such a prime factor or not. We can also ask if it is possible to
have more than one primitive prime divisor and how big these primes are. These are
the questions we intend to answer here.

6.1   The History of Primitive Divisors

Recall that in an integer sequence (a), a prime p is a *primitive* divisor of a_n if $p \mid a_n$ but
$p \nmid a_k$ for $k < n$. In this chapter, we will consider the primitive divisors of Fibonacci
sequence, the Lucas sequence of first kind, and the Lehmer sequence $(\bar{U})$. For brevity,
we will use (U) to denote the associated Lehmer sequence $(\bar{U})$ that is always integer.
In doing so, we will refine the definition of primitive divisor for Lucas and Lehmer
sequence. We will not be discussing Lucas and Lehmer sequences in the same section,
so it will be alright to denote them with (U) since the context will be clear. p is a
primitive divisor of the Lucas number U_n if $p \mid U_n$ but $p \nmid (\alpha - \beta)^2 U_2 \cdots U_{n-1}$. In

© The Author(s), under exclusive license to Springer Nature Singapore Pte Ltd. 2021   119
M. Billal and S. Riasat, *Integer Sequences*,
https://doi.org/10.1007/978-981-16-0570-3_6

a similar fashion, p is a primitive divisor of the Lehmer number U_n if $p \mid U_n$ but $p \nmid (\alpha^2 - \beta^2)^2 U_2 \cdots U_{n-1}$. Moreover, in Chap. 5, we considered that $a > 0$, a is real and $\Delta > 0$. In this case, we have α and β are real roots. We call a Lehmer sequence (U) *real* if α, β are real. Otherwise, (U) is *complex*. The same terminology applies for Lucas sequence as well. Also, a primitive prime divisor p of U_n is *intrinsic* if $p \mid n$, otherwise p is *extrinsic*. If a Lucas or Lehmer number U_n does not have any primitive prime divisor, then U_n is *exceptional* and n is an *exceptional index*. Many authors use the term *intrinsic* in place of *primitive*. However, we reserve primitive to mean any prime divisor that has not appeared in the sequence before regardless of whether it divides the index or not. We say p is intrinsic (extrinsic) *if* p is a primitive divisor of U_n and $p \mid n$ ($p \nmid n$). Note that, V_n has a primitive prime factor if and only if U_{2n} has a primitive factor due to the fact $U_{2n} = U_n V_n$. It is easy to see because p cannot divide both U_n and V_n.

In the literature of primitive prime divisors, the most crucial result is probably Bang's theorem. Bang [Ban86] proved that $2^n - 1$ has a primitive prime divisor except when $n = 6$. Then Zsigmondy [Zsi92] generalized this result for two relatively prime integers a and b, which states that $a^n - b^n$ has a primitive prime divisor except when $a = 2, b = 1, n = 6$ or $a + b = 2^k, n = 2$. Similarly, $a^n + b^n$ has a primitive prime divisor except when $a = 2, b = 1, n = 3$. Birkhoff and Vandiver [BV04] also proved this result and showed that the greatest prime factor of $a^n - b^n$ is at least $n + 1$ for $n > 2$. Carmichael [Car13] extended this for real Lucas sequences. He proved that if $n \notin \{1, 2, 6\}$, then the Lucas number U_n has at least one primitive divisor except when $n = 12, a = 1, b = -1$. Almost a 100 years later, Yabuta [Yab01] proved the same result but in a more elegant way. Lehmer [Leh30, Page 432] himself discussed intrinsic divisors of (U). Ward [War55] extended Carmichael's result for real Lehmer sequence under the condition that $(\alpha + \beta)^2$ and $\alpha\beta$ are integers. Ward also showed that the greatest prime factor of U_n is greater than or equal to $n - 1$ for $n \geq 19$. Durst [Dur59], [Dur61] showed that the same holds true for $n > 12$. The same techniques they use do not apply to complex Lehmer sequences. At the time Ward wrote his paper, no progress was made for primitive divisors of complex Lucas or Lehmer sequences. Carmichael [Car13, Page 52] showed using some simple examples that complex Lucas or Lehmer sequences may have many U_n with no primitive divisors.

Schinzel [Sch62b] proved for the first time that a complex Lehmer number U_n has a primitive prime divisor for all $n > n_0$, where n_0 is a constant dependent on α, β that can be effectively computed. His short proof relies on a result of Gelfond [Gel60, Theorem IV, Page 174] on linear forms. Schinzel calls this result of Gelfond a *deep theorem*. Schinzel and Postnikova [L P74] removed the dependency of this constant on α, β using a result of Baker [Bak72, Bak73]. Stewart [Ste77] proved that $e^{452} 4^{67}$ is a value of this constant. He also showed that n_0 is 6 and 12 for Lucas and Lehmer sequences, respectively, with some exceptions similar to the ones we mentioned for Zsigmondy's theorem. These exceptions could be found by solving some Thue equations. Voutier [Vou95, Vou96, Vou98] improved the value of n_0 to 30030. Bilu et al. [BHV01] solved the equations in Stewart [Ste77] and finally showed that the Lucas or Lehmer number U_n has a primitive divisor for $n > 30$.

Schinzel [Sch62a] proves a result regarding $a^n - b^n$ having two primitive prime factors. Rotkiewicz [Rot62] proves a similar result for Lucas sequences under the assumption $a, b > 0$. Schinzel [Sch63a] again generalizes this result for Lehmer sequences. His result is that U_n has at least two primitive prime factors except some cases under the condition that $n \neq 1, 2, 3, 4, 6$, $\chi = \text{rad}(b \max(c, \Delta))$, ϱ is 1 or 2 based on whether $\chi \equiv 1 \pmod 4$ or $\chi \equiv 2, 3 \pmod 4$ respectively and $\frac{n}{\varrho \chi}$ is an odd integer. Schinzel [Sch63b] extends this result under some conditions and shows that U_n may have 3, 4 or 6 primitive prime divisors. Schinzel [Sch68] proves similar results for complex Lehmer numbers. We omit the actual statements because they are too long and adds little to our discussion.

6.2 Primitive Divisors of $a^n - b^n$

Birkhoff and Vandiver [BV04], Zsigmondy [Zsi92] both discuss the primitive divisors of $a^n - b^n$ for relatively prime positive integers a, b. We proved in (4.3) that

Theorem 6.1 *Let a and b be two relatively prime positive integers. Then*

$$\gcd\left(\frac{a^n - b^n}{a - b}, a - b\right) = \gcd(a - b, n)$$

holds true for any positive integer n.

So, if n is relatively prime to $a - b$, then $\frac{a^n - b^n}{a - b}$ is relatively prime to $a - b$. Billal [Bil12] discusses this property and shows some applications in some problems. From the theory we have established in Chap. 5, we can say that $a^n - b^n$ has the exponent lifting property. We can consider the cyclotomic polynomials of $a^n - b^n$ and simply conclude that. But here, we will show a proof that uses more innate properties of $a^n - b^n$. This property has also been discussed in detail. For example, see the article *a nice and tricky lemma* in Andreescu [And14, Chap. 3, Sect. 3.25, Page 535–540].

Theorem 6.2 (Lifting the exponent lemma) *Let p be and odd prime and x, y be integers such that none of them are divisible by p and $p \mid x - y$. Then*

$$\nu_p(x^n - y^n) = \nu_p(x - y) + \nu_p(n)$$

holds true for any positive integer n. If $p = 2$ and $2 \mid x - y$, then

$$\nu_2(x^n - y^n) = \begin{cases} \nu_2(x - y) + \nu_2(n) & \text{if } 4 \mid x - y \\ \nu_2(x - y) + \nu_2(x + y) + \nu_2(n) - 1 & \text{if } 2 \mid n \end{cases}$$

The following results are necessary for proving the theorem.

Theorem 6.3 *Let p be an odd prime divisor of $a - b$. If $p^\alpha \parallel a - b$, then $p^{\alpha+1} \parallel a^p - b^p$.*

Proof Set $\frac{a-b}{p^\alpha} = m$ so that $p \nmid m, a = p^\alpha m + b$.

$$a^p = (b + p^\alpha m)^p$$

$$= b^p + \sum_{i=1}^{p-1} \binom{p}{i} b^{p-i} (p^\alpha m)^i + p^{p\alpha} m^p$$

$$a^p - b^p = \sum_{i=1}^{p-1} p \binom{p-1}{i-1} b^{p-i} p^{i\alpha} m^i + p^{p\alpha} m^p$$

$$= p^{\alpha+1} \left(\binom{p-1}{0} b^{p-1} m^i + \binom{p-1}{1} b^{p-2} m^2 p^\alpha + \cdots + p^{\alpha(p-1)-1} m^p \right)$$

The last line follows from the facts that $\binom{n}{k} = \frac{n}{k} \binom{n-1}{k-1}$ and $p \mid \binom{p}{i}$ for $0 < i < p$. Thus, we get that $a^p - b^p$ is divisible by $p^{\alpha+1}$ but since the other term is not divisible by p, $a^p - b^p$ is not divisible by a higher power of p. ∎

Letting $n = p^r k$ with $p \nmid k$ and combining this result with Theorem 6.1, we get Theorem 6.2. Now, assume that the product of all primitive divisors of $a^n - b^n$ is $\chi(n)$. If $p_r, p_s, \cdots$ are the primitive prime divisors of $a^n - b^n$, then

$$\chi(n) = p_r^{e_r} p_s^{e_s} \cdots$$

Let us consider the cyclotomic polynomials of $a^n - b^n$ and recall some results. If $\Phi_m(a, b)$ and $\Phi_n(a, b)$ share a common factor other than 1, then $\frac{m}{n} = p^u$ for some non-negative u. Consequently, we also proved that if p is an odd prime factor of $\Phi_n(x)$ and $p \mid \Phi_d(x)$ for some $d \mid n$ with $d < n$, then $p \mid n$. Thus, the factor $\Phi_n(a, b)$ contains the all primitive factors of $a^n - b^n$ as well as this kind of p such that $p \mid n$. However, we can easily remove such factors from $\Phi_n(a, b)$ so that $\Phi_n(a, b)$ can be considered what [BV04, Page 176] calls the *algebraic primitive factor* of $a^n - b^n$. Next, we will try to determine the ratio $\Psi_n = \frac{\Phi_n(a,b)}{\chi_n}$. Here, Ψ_n is the product of primes p such that $p \mid \Phi_n(a, b)$ and $n = p^\beta m$ with $\gcd(p, m) = 1$. By Theorem 1.39, we have

$$\Phi_n(a, b) = \frac{\Phi_m(a^{p^\beta}, b^{p^\beta})}{\Phi_m(a^{p^{\beta-1}}, b^{p^{\beta-1}})}$$

Applying Theorem 1.49 for $\Phi_n(a, b)$ and using the fact that $\Phi_n(x)$ is an integer polynomial, we get

$$\Phi_m\left(a^{p^\beta}, b^{p^\beta}\right) \equiv \Phi_m(a, b)^{p^\beta} \pmod{p}$$

$$\Phi_n(a, b) \equiv \frac{\Phi_m\left(a^{p^\beta}, b^{p^\beta}\right)}{\Phi_m\left(a^{p^{\beta-1}}, b^{p^{\beta-1}}\right)} \pmod{p}$$

$$\equiv \frac{\Phi_m(a, b)^{p^\beta}}{\Phi_m(a, b)^{p^{\beta-1}}} \pmod{p}$$

$$\equiv \Phi_m(a, b)^{p^\beta - p^{\beta-1}} \pmod{p}$$

$$\equiv \Phi_m(a, b)^{p^{\beta-1}(p-1)} \pmod{p}$$

So, p divides $\Phi_n(a, b)$ if and only if p divides $\Phi_m(a, b)$. If p does not divide $\Phi_m(a, b)$ for any prime divisor p of n, then $\Phi_n(a, b) = \chi_n$. On the other hand, if there is such a p, then $\gcd(\Psi_m, p) = 1$ where $\Psi_m = \frac{\Phi_m(a,b)}{\chi_m}$. In order for this to hold, p must divide χ_m. Since p divides $\Phi_m(a, b)$ and $p \nmid m$, by Theorem 1.45, p must be 1 (mod m). This implies that $p > m = \frac{n}{p^\beta}$. Thus, p is the largest prime divisor of n and as a consequence, $\Psi_n = p^\lambda$ for some positive integer λ.

Let ρ be the smallest positive integer i such that $p \mid a^i - b^i$. Then, $p \mid \Phi_\rho(a, b)$ and $p \mid \Phi_m(a, b)$. Again, we have that $\frac{m}{\rho} = p^s$ for some non-negative integer s. Assume that $p^\omega \parallel \Phi_\rho(a, b)$, then by Theorem 6.2, $p^{\omega+s} \parallel a^m - b^m$ and $p^{\omega+s\beta} \parallel a^n - b^n$. Since $n = \rho p^{s+\beta}$, we have the factorization

$$a^n - b^n = \prod_{d \mid n} \Phi_d(a, b)$$

$$= \left(\prod_{d \mid p^{s+\beta}} \Phi_{\rho d}(a, b)\right) \cdot \left(\prod_{\substack{d \mid \rho \\ d < \rho \\ 0 \le i \le s+\beta}} \Phi_{d p^i}(a, b)\right)$$

Clearly, p does not divide $\Phi_{d p^i}(a, b)$ if $d < \rho$. Since $p \mid \Phi_{\rho d}$ for any divisor d of $p^{s+\beta}$, if $p^2 \mid \Phi_{\rho d}$ for any divisor $d > 1$ of $p^{s+\beta}$, then we get

$$\nu_p(a^n - b^n) = \sum_{d \mid p^{s+\beta}} \nu_p(\Phi_{\rho d}(a, b))$$

$$= \nu_p(\Phi_\rho(a, b)) + \sum_{i=1}^{s+\beta} \nu_p(\Phi_{\rho p^i}(a, b))$$

$$> \omega + s + \beta$$

This is impossible, so we must have that $p \parallel \Phi_{\rho d}(a, b)$ for $d \mid p^{s+\beta}$ if $d > 1$. Thus, we have the following result.

Theorem 6.4 *If $n \neq 2$, then*

$$\Psi_n = \begin{cases} p & \text{if } p \mid \Phi_m(a,b) \\ 1 & \text{otherwise} \end{cases}$$

where $m = \frac{n}{p^\beta}$ and $p \nmid m$.

A corollary of this result is that if $\chi_n > 1$, then $a^n - b^n$ has a primitive divisor. As in Theorem 6.4, we have two cases to check. First we check the case $\Psi_n = p$, so $\chi_n = \frac{\Phi_n(a,b)}{p}$ where $n = p^\beta m$ with $p \nmid m$ and $p \equiv 1 \pmod{m}$ such that $p \mid \Phi_m(a,b)$. As $\Phi_m(a,b)$ cannot be 0, $\Phi_m(a,b) \geq p$ so

$$\chi_n \geq \frac{\Phi_n(a,b)}{\Phi_m(a,b)}$$

From Theorem 3.12,

$$\Phi_n(a,b) = \frac{(a^n - b^n) \prod \left(a^{\frac{n}{pq}} - b^{\frac{n}{pq}}\right) \prod \left(a^{\frac{n}{pqrs}} - b^{\frac{n}{pqrs}}\right) \cdots}{\prod \left(a^{\frac{n}{p}} - b^{\frac{n}{p}}\right) \prod \left(a^{\frac{n}{pqr}} - b^{\frac{n}{pqr}}\right) \cdots} \tag{6.1}$$

Let us take $m > 1$ and $a > b$.

$$a^n - b^n = (a - b)(a^{n-1} + a^{n-2}b + \cdots + b^{n-1})$$
$$\geq a^{n-1} + (n-1)b^{n-1}$$
$$> a^{n-1}$$

Using this in the numerator of (6.1),

$$(a^n - b^n)\prod\left(a^{\frac{n}{pq}} - b^{\frac{n}{pq}}\right)\prod\left(a^{\frac{n}{pqrs}} - b^{\frac{n}{pqrs}}\right)\cdots > a^{n-1}\prod\left(a^{\frac{n}{pq}} - 1\right)\prod\left(a^{\frac{n}{pqrs}} - 1\right)\cdots$$
$$> a^{n-1+\sum\left(\frac{n}{pq}-1\right)+\sum\left(\frac{n}{pqrs}-1\right)\cdots}$$
$$> a^{n+\sum\frac{n}{pq}+\sum\frac{n}{pqrs}+\cdots-2^{k-1}}$$

where k is the number of distinct prime factors of n. Again,

$$a^n - b^n < a^n$$

Using this in the denominator,

$$\prod\left(a^{\frac{n}{p}} - b^{\frac{n}{p}}\right)\prod\left(a^{\frac{n}{pqr}} - b^{\frac{n}{pqr}}\right)\cdots < a^{\sum\frac{n}{p}+\sum\frac{n}{pqr}+\cdots}$$

Combining the results for numerator and denominator,

$$\Phi_n(a,b) > \frac{a^{n+\sum \frac{n}{pq}+\sum \frac{n}{pqrs}+\cdots-2^{k-1}}}{a^{\sum \frac{n}{p}+\sum \frac{n}{pqr}+\cdots}}$$

$$> a^{n-\sum \frac{n}{p}+\sum \frac{n}{pq}-\sum \frac{n}{pqr}+\cdots-2^{k-1}}$$

by principle of inclusion and exclusion, $n - \sum \frac{n}{p} + \sum \frac{n}{pq} - \sum \frac{n}{pqr} + \cdots = \varphi(n)$ so we get

$$\Phi_n(a,b) > a^{\varphi(n)-2^{k-1}}$$

Again, using $a^n - b^n < a^n$ in the numerator of $\Phi_m(a,b)$ and $a^n - b^n > a^{n-1}$ in the denominator,

$$\Phi_m(a,b) > a^{\varphi(m)-2^{k-2}}$$

since m has exactly one less prime factor than n. Combining this with the inequality for $\Phi_n(a,b)$,

$$\chi_n \geq \frac{\Phi_n(a,b)}{\Phi_m(a,b)}$$

$$> \frac{a^{\varphi(n)-2^{k-1}}}{a^{\varphi(m)-2^{k-2}}}$$

$$> a^{\varphi(n)-\varphi(m)-3\cdot 2^{k-2}}$$

$\varphi(n)$ is a multiplicative function, so using $n = p^\beta m$, $\varphi(n) = \varphi(p^\beta)\varphi(m)$. By assumption, $m > 1$ so $\varphi(m) \geq 2^{k-2}$. We want to show that the exponent of a is at least 0 in the inequality above. Now, $\varphi(p^\beta) = p^{\beta-1}(p-1)$ and for $p > 3$,

$$p^{\beta-1}(p-1) - 1 \geq 3$$

Using these, we get that

$$\varphi(m)(p^{\beta-1}(p-1)-1) \geq 3 \cdot 2^{k-2}$$

$$\varphi(n) - \varphi(m) - 3 \cdot 2^{k-2} \geq 0$$

holds except when $p = 3$. In the case $p = 3$, we have $n = 6$. Now, $\Phi_6(a,b) = a^2 - ab + b^2 > 3$ except when $a = 2$, $b = 1$. On the other case, $m = 1$ and $n = p^\beta$.

$$\Phi_{p^\beta} = \frac{a^{p^\beta} - b^{p^\beta}}{a^{p^{\beta-1}} - b^{p^{\beta-1}}}$$

$$= \left(a^{p^{\beta-1}}\right)^{p-1} + \cdots + \left(b^{p^{\beta-1}}\right)^{p-1}$$

$$> p$$

$$\chi_n > 1$$

We are left with the case $\chi_n = 1$. In this case, $\Phi_n(a, b) = \chi_n$ and $\Phi_n(a, b) > a^{\varphi(n)-2^{k-1}}$. Again, using $\varphi(n) \geq 2^{k-1}$, we easily get $\Phi_n(a, b) > 1$. Thus, we have proven the following theorem.

Theorem 6.5 (Zsigmondy's theorem) *For a positive integer $n > 1$, $a^n - b^n$ has a primitive prime divisor except the following cases:*

$$n = \begin{cases} 1 & and\ a - b = 1 \\ 2 & and\ a + b = 2^k \\ 6 & and\ a = 2, b = 1 \end{cases}$$

And $a^n + b^n$ has a primitive divisor except when $a = 2, b = 1, n = 3$.

We also have the following result as a corollary.

Theorem 6.6 *Let $n > 2$ be an integer. Then $a^n - b^n$ has at least one prime divisor $1 \pmod{n}$.*

Let $P(n)$ be the largest prime factor of n. Then the next theorem is implicated by the result above.

Theorem 6.7 *Let $n > 2$ be an integer. Then $P(a^n - b^n) \geq n + 1$.*

We have the next result inspired by the proof of lifting the exponent lemma.

Theorem 6.8 (Exponent lifting criteria) *Let (a_n) be a strong divisibility sequence, (b_n) be the* lcm *sequence of (a_n), and ρ be the rank of apparition of prime p in (a_n). Then (a_n) has the exponent lifting property if and only if for any positive integers n and $m > 1$ such that $p \nmid m$, $p \parallel b_{\rho p^n}$ but $p \nmid b_{\rho p^n m}$.*

Proof First, we will prove the if part. Since (a_n) is a strong divisibility sequence, $p \mid a_k$ if and only if $\rho \mid k$. By assumption, (a_n) has exponent lifting property. If $p^\alpha \parallel a_\rho$, then $p^{\alpha+1} \parallel a_{\rho p}$.

$$a_{\rho p} = \prod_{d \mid \rho p} b_d$$

$$\nu_p(a_{\rho p}) = \nu_p \left(\prod_{d \mid \rho p} b_d \right)$$

If $d < \rho$, then $p \nmid a_d$ so $p \nmid b_d$. Thus,

$$\nu_p(a_{\rho p}) = \nu_p\left(\prod_{d|p} b_{\rho d}\right)$$
$$= \nu_p(b_\rho) + \nu_p(b_{\rho p})$$
$$\alpha + 1 = \alpha + \nu_p(b_{\rho p})$$

So, $\nu_p(a_{\rho p}) = 1$ and $p \mid b_{\rho p}$. For a positive integer $i > 1$, we have

$$a_{\rho p^i} = \prod_{d|\rho p^i} b_d$$
$$= \prod_{d|\rho p} b_d \prod_{\substack{d|\rho p^i \\ d \nmid \rho p}} b_d$$
$$= a_{\rho p} \prod_{\substack{d|\rho p^i \\ d \nmid \rho p}} b_d$$

By induction, we can see that p not only divides $b_{\rho p^i}$ for $i \in \mathbf{N}$, more precisely, $p \parallel b_{\rho p^i}$. Next, assume that $p^{\alpha+u} \parallel a_n$ for some positive integer $n = \rho p^u m$ where $p \nmid m$. From the exponent lifting property,

$$\nu_p(a_n) = \nu_p(a_{\rho p^u m})$$
$$= \nu_p(a_\rho) + \nu_p\left(\prod_{d|p^u m} b_{\rho d}\right)$$
$$= \alpha + \nu_p\left(\prod_{d|p^u} b_{\rho d}\right) + \nu_p\left(\prod_{\substack{d|p^u \\ e|m \\ e>1}} b_{\rho d e}\right)$$
$$\alpha + u = \alpha + \sum_{i=1}^{u} \nu_p(b_{\rho p^i}) + \nu_p\left(\prod_{i=1}^{u}\prod_{\substack{e|m \\ e>1}} b_{\rho p^i e}\right)$$
$$= \alpha + u + \nu_p\left(\prod_{i=1}^{u}\prod_{\substack{e|m \\ e>1}} b_{\rho p^i e}\right)$$
$$= \alpha + u + \sum_{i=1}^{u}\sum_{\substack{e|m \\ e>1}} \nu_p(b_{\rho p^i e})$$

From this, we have that $\nu_p(b_{\rho p^i e}) = 0$ for $1 \le i \le u$ and $e \mid m$ if $e > 1$. In other words, $p \mid b_k$ if and only if $k = \rho p^u$ for some non-negative integer u. So, $p \mid b_m$ and $p \mid b_n$ with $m \ge n$ if and only if $\frac{m}{n} = p^v$ for some non-negative integer v.

For the only if part, we have that (a_n) is a strong divisibility sequence such that $p \parallel b_{\rho p^u}$ but $p \nmid b_{\rho p^u m}$ for $m > 1$. Let n be a positive integer such that $n = \rho p^u m$ and $p^\alpha \parallel a_\rho$.

$$\nu_p(a_n) = \nu_p(a_{\rho p^u m})$$

$$= \nu_p \left(\prod_{d \mid \rho p^u m} b_d \right)$$

$$= \nu_p(a_\rho) + \nu_p \left(\prod_{d \mid p^u m} b_{\rho d} \right)$$

$$= \nu_p(a_\rho) + \sum_{d \mid p^u} \nu_p(b_{\rho d}) + \sum_{d \mid p^u} \sum_{\substack{e \mid m \\ e > 1}} \nu_p(b_{\rho d e})$$

$$= \alpha + \sum_{i=1}^{u} \nu_p(b_{\rho p^i}) + 0$$

$$= \alpha + \sum_{i=1}^{u} 1$$

$$= \alpha + u$$

This proves the theorem.   ∎

6.3   Primitive Divisors of Real Lucas Sequences

Recall that a Lucas sequence of the first kind is of the form

$$U_n = \frac{\alpha^n - \beta^n}{\alpha - \beta}$$

where α, β are the roots of the equation $x^2 - ax + b = 0$. Carmichael [Car13, Theorem XXIII, Page 61] proved the following theorem.

Theorem 6.9 *If (U) is a real Lucas sequence, then U_n contains a primitive prime divisor except when $n = 12, \alpha + \beta = \pm 1, \alpha \beta = -1$.*

The proof is quite long, and they use the results we developed in Chap. 5. We will show the proof by Yabuta [Yab01] because it is quite elegant and elementary. We will again make use of some results we have already established.

Theorem 6.10 *Let $n \neq 1, 2, 6$ be a positive integer and the prime factorization of n be $n = p_1^{e_1} p_2^{e_2} \cdots p_r^{e_r}$. A sufficient condition that U_n has a primitive prime divisor is*

$$|\Phi_n(\alpha, \beta)| > p_1 p_2 \cdots p_r$$

Proof Assume that U_n has no primitive prime divisor. If $p \mid \Phi_d(\alpha, \beta)$ for some $0 < d < n$, then $p \mid n$ and $p^2 \nmid \Phi_n(\alpha, \beta)$. Thus $\Phi_n(\alpha, \beta)$ must divide the product $p_1 p_2 \cdots p_r$. The result now follows by contraposition. ∎

Let $n > 2$ be a positive integer and $a^2 - 4b$ be positive.

$$\Phi_n(\alpha, \beta) = \prod_{\substack{i=1 \\ \gcd(i,n)=1}} \left(\alpha - \zeta^i \beta\right)$$

$$= \prod_{\substack{i=1 \\ \gcd(i,n)=1 \\ i < \frac{n}{2}}} \left(\alpha - \zeta^i \beta\right)\left(\alpha - \zeta^{-i}\beta\right)$$

$$= \prod_{\substack{i=1 \\ \gcd(i,n)=1 \\ i < \frac{n}{2}}} \left((\alpha + \beta)^2 - \alpha\beta(2 + \zeta^i + \zeta^{-i})\right)$$

where $\zeta = e^{\frac{2i\pi}{n}}$. Setting $2 + \zeta^i + \zeta^{-i} = \gamma_i$,

$$\Phi_n(\alpha, \beta) = \prod_{\substack{i=1 \\ \gcd(i,n)=1 \\ i < \frac{n}{2}}} \left(a^2 - b\gamma_i\right)$$

We want to achieve a minimum value of $\Phi_n(\alpha, \beta)$.

Theorem 6.11 *Let $n > 2$ be an integer. Then $\Phi_n(\alpha, \beta)$ is minimum if $(a, b) \in \{(1, -1), (3, 2)\}$.*

Proof For a particular j, we have

$$\gamma_j = 2 + \zeta^i + \zeta^{-i}$$
$$= 2 + e^{\frac{2i\pi j}{n}} + e^{-\frac{2i\pi j}{n}}$$
$$= 2 + \cos\frac{2\pi j}{n} + i \sin\frac{2\pi j}{n} + \cos\frac{2\pi j}{n} - i \sin\frac{2\pi j}{n}$$
$$= 2\left(1 + \cos\frac{2\pi j}{n}\right)$$
$$= 4\cos^2\frac{\pi j}{n}$$

Since $n > 2, 0 < \gamma_j < 4$. So, if $b < 0$, then

$$a^2 - b\gamma_j \geq 1 + \gamma_j$$

where equality holds if and only if $a = 1, b = -1$. If $b > 0$, then consider two cases: $b = 1$ and $b > 1$. Since α, β are real and unequal, by AM-GM inequality, $\alpha + \beta > 2\sqrt{\alpha\beta}$ so $a \geq 3$.

$$a^2 - b\gamma_j \geq 9 - \gamma_j$$
$$> 9 - 2\gamma_j$$

If $b > 1$,

$$a^2 - 4b \geq 1$$
$$a^2 - b\gamma_j \geq 1 + 4b - b\gamma_j$$
$$\geq 9 - 2\gamma_j + (b-2)(4-\gamma_j)$$
$$\geq 9 - 2\gamma_j$$

Equality occurs if and only if $b = 2, a = 3$.   ∎

Thus, we have the following result.

Theorem 6.12 *Let $n \neq 1, 2, 6$ be a positive integer. If $\Phi_n(\alpha, \beta)$ is greater than the product of distinct prime factors of n where α, β are roots of either $x^2 - x - 1$ or $x^2 - 3x + 2$, then a real Lucas number U_n has a primitive prime divisor.*

Yabuta calls the Lucas sequence generated by the roots of $x^2 - x - 1 = 0$ the *Fibonacci sequence* and the sequence generated by the roots of $x^2 - 3x + 2 = 0$ the *Fermat sequence*. The theorem above is very useful for proving Carmichael's theorem because we only have to check the case for Fibonacci and Fermat sequences.

Theorem 6.13 *If $n > 2$ and $|x| < \frac{1}{2}$,*

$$\Phi_n(x) \geq 1 - |x| - x^2$$

Proof Using Theorem 3.11 Yabuta rewrites $\Phi_n(x)$ as

$$\Phi_n(x) = \prod_{d|n} \left(1 - x^{\frac{n}{d}}\right)^{\mu(d)}$$

Now, if $\mu(d) = 1$, then $(1 - x^{\frac{n}{d}}) \geq (1 - |x|^{\frac{n}{d}})$. If $\mu(d) = 0$, then $(1 - |x|^{\frac{n}{d}})$ is definitely smaller than 1. And if $\mu(d) = -1$, then $(1 - x^{\frac{n}{d}})(1 - |x|^{\frac{n}{d}})$ is at most 1 if $|x| < \frac{1}{2}$. We also use $(1 - x)(1 - y) \geq 1 - x - y$ if $0 \leq x, y \leq 1$.

$$(1 - x^{\frac{n}{d}})^{\mu(d)} \geq (1 - |x|^{\frac{n}{d}})$$

$$\Phi_n(x) = \prod_{d|n} (1 - x^{\frac{n}{d}})^{\mu(d)}$$

$$\geq \prod_{i \geq 1} (1 - |x|)^i$$

$$\geq (1 - |x|)(1 - |x| - |x|^2 - \cdots)$$

$$\geq (1 - |x|) \left(1 - \frac{|x|^2}{1 - |x|} \right)$$

$$\geq 1 - |x| - |x|^2$$

∎

Theorem 6.14 *If $n \neq 1, 2, 6, 12$, then F_n contains at least one primitive divisor.*

Proof The characteristic polynomial of (F) is $x^2 - x - 1 = 0$ and it has roots $\alpha = \frac{1+\sqrt{5}}{2}, \beta = \frac{1-\sqrt{5}}{2}$.

$$\frac{\beta}{\alpha} = \frac{3 - \sqrt{5}}{2}$$

$$< 2$$

$$\Phi_n \left(\frac{\beta}{\alpha} \right) \geq 1 - \left| \frac{\beta}{\alpha} \right| - \left| \frac{\beta}{\alpha} \right|^2$$

$$\geq 2\sqrt{5} - 4$$

$$> \frac{2}{5}$$

Using $\alpha > \frac{3}{2}$, we get

$$\Phi_n(\alpha, \beta) = \alpha^{\varphi(n)} \Phi_n \left(\frac{\beta}{\alpha} \right)$$

$$> \left(\frac{3}{2} \right)^{\varphi(n)} \frac{2}{5}$$

$$\varphi(n) = p_1^{e_1-1}(p_1 - 1) p_2^{e_2-1}(p_2 - 1) \cdots p_r^{e_r-1}(p_r - 1)$$

Without loss of generality, assume that $p_1 \leq p_2 \leq \cdots p_r$. Now, if $p_1 > 7$,

$$3^{p_1-1} > 5 \cdot 2^{p_1-1}$$

$$\frac{2}{5} \left(\frac{3}{2} \right)^{p_1-1} > p_1$$

For $i > 1$, $p_i > p_1 > 11$ and using $x^{m-1} > my$ for $x > y$, $\left(\frac{3}{2} \right)^{p_i-1} > p_i$.

$$\Phi_n(\alpha, \beta) > \left(\frac{3}{2}\right)^{\varphi(n)} \frac{2}{5}$$

$$> \left(\frac{2}{5}\right)\left(\frac{3}{2}\right)^{p_1-1} \cdots \left(\frac{3}{2}\right)^{p_r-1}$$

$$> p_1 p_2 \cdots p_r$$

Thus, p_1 cannot be greater than 7 and $n = 2^a 3^b 5^c 7^d$ where $0 \le a \le 3, 0 \le b \le 2, 0 \le c \le 1, 0 \le d \le 1$. We can check that

$$\Phi_2(\alpha, \beta) = 1$$
$$\Phi_3(\alpha, \beta) = 2$$
$$\Phi_4(\alpha, \beta) = 3$$
$$\Phi_5(\alpha, \beta) = 5$$
$$\Phi_6(\alpha, \beta) = 4$$
$$\Phi_7(\alpha, \beta) = 13$$
$$\Phi_8(\alpha, \beta) = 7$$
$$\Phi_9(\alpha, \beta) = 17$$
$$\Phi_{10}(\alpha, \beta) = 11$$
$$\Phi_{12}(\alpha, \beta) = 6$$
$$\Phi_{14}(\alpha, \beta) = 29$$
$$\Phi_{15}(\alpha, \beta) = 61$$
$$\Phi_{18}(\alpha, \beta) = 19$$
$$\Phi_{30}(\alpha, \beta) = 31$$

So, $\Phi_n(\alpha, \beta) > p_1 p_2 \cdots p_r$ holds for $n \ne 1, 2, 3, 5, 6, 12$. However, $F_3 = 2$ and $F_5 = 5$ so F_n has primitive divisor for $n = 3, 5$ while others do not.   ∎

Theorem 6.15 *If $n \ne 1, 2, 6$, then n term of Fermat sequence has a primitive divisor.*

Proof The roots of characteristic equation $x^2 - 3x + 2 = 0$ is $\alpha = 2, \beta = 1$.

$$\Phi_n\left(\frac{\beta}{\alpha}\right) \ge 1 - \left|\frac{\beta}{\alpha}\right| - \left|\frac{\beta}{\alpha}\right|^2$$

$$\ge \frac{1}{4}$$

$$\Phi_n(\alpha, \beta) \ge 2^{\varphi(n)} \frac{1}{4}$$

For all $n > 2$, $2^{\varphi(n)} \frac{1}{4} > \frac{2}{5} \left(\frac{3}{2}\right)^{\varphi(n)}$. In a similar fashion as the previous proof, we can verify the claim.   ∎

Finally, we can finish the theorem. If $n \neq 1, 2, 6, 12$, then $\Phi_n(\alpha, \beta)$ is greater than $p_1 p_2 \cdots p_r$. Thus both Fermat and Fibonacci numbers F_n have a primitive divisor by the results above. Since $\Phi_n(\alpha, \beta)$ is minimum for Fermat and Fibonacci sequences only and both of them have $\Phi_n(\alpha, \beta) > p_1 p_2 \cdots p_r$, the result holds. Moreover, $\Phi_3(\alpha, \beta) = a^2 - ab + b^2 = (a - b)^2 + ab > 3$ except when $a = 1, b = -1$. Also, $\Phi_5(\alpha, \beta) > 5$ and $\Phi_{12}(\alpha, \beta) > 6$ except for Fibonacci sequence. Thus, if $n \neq 1, 2, 6$, then U_n has a primitive divisor except when $a = 1, b = -1$ and $n = 12$.

6.4 Primitive Divisors of Real Lehmer Sequences

As we mentioned earlier, Ward [War55] proved that Lehmer sequences have primitive divisors. In this section, we will prove that (U) defined as

$$U_n = \begin{cases} \dfrac{\alpha^n - \beta^n}{\alpha - \beta} & \text{if } n \text{ is odd} \\ \dfrac{\alpha^n - \beta^n}{\alpha^2 - \beta^2} & \text{otherwise} \end{cases}$$

where α, β are real roots of the equation $x^2 - ax + b = 0$ such that $(\alpha + \beta)^2$ and $\alpha\beta$ are integers. Let $\mathcal{D} = a - 4b$ and R as

$$R = \begin{cases} |4ab| & \text{if } m \text{ is negative} \\ |4\mathcal{D}b| & \text{if } m \text{ is positive} \end{cases}$$

The primary result is the theorem below.

Theorem 6.16 *A real Lehmer sequence (U) is exceptional only if $R < 16$. If $n > 18$, then U_n always has a primitive divisor.*

As usual, we assume without loss of generality that $\gcd(a, b) = 1$ and $a > 0$. We will recall the divisibility properties we proved in Chap. 5, especially the properties of cyclotomic divisors of Lehmer numbers in Sect. 5.5. We proved similar results in Sect. 6.2 of this chapter. Here is a list of the results we will require.

(1) If ρ is the rank of apparition of an odd prime p in (U), then $p \mid U_n$ if and only if $\rho \mid n$.
(2) If $p^\alpha \| U_\rho$, then $p^{\alpha+\beta} \| U_{\rho m p^\beta}$ where $p \nmid m$.
(3) For a positive integer k, $p \| \Phi_{p^k \rho}(\alpha, \beta)$.
(4) If $p \mid U_n$ where $n = \rho k p^\alpha$ for some $\alpha \geq 0$ and $p \nmid k$,, then $p \nmid \Phi_n(\alpha, \beta)$.
(5) If $p \mid \Phi_n(\alpha, \beta)$ and $p \mid \Phi_d(\alpha, \beta)$ for some $0 < d < n$, then $p \| \Phi_n(\alpha, \beta)$.
(6) A sufficient condition that U_n has a primitive divisor is that $|\Phi_n(\alpha, \beta)| > n$.

In the same fashion as previous section, we have

$$\Phi_n(\alpha, \beta)^2 = \prod_{\substack{1 \le j \le n \\ \gcd(j,n)=1}} \left(\alpha - e^{\frac{2i\pi j}{n}} \beta \right) \cdot \prod_{\substack{1 \le j \le n \\ \gcd(j,n)=1}} \left(\alpha - e^{-\frac{2i\pi j}{n}} \beta \right)$$

$$= \prod_{\substack{1 \le j \le n \\ \gcd(j,n)=1}} \left(\alpha^2 + \beta^2 - \alpha\beta \left(e^{\frac{2i\pi j}{n}} + e^{-\frac{2i\pi j}{n}} \right) \right)$$

$$= \prod_{\substack{1 \le j \le n \\ \gcd(j,n)=1}} \left(a - 4b \cos^2 \frac{\pi j}{n} \right)$$

$$= \prod_{\substack{1 \le j \le n \\ \gcd(j,n)=1}} \left(\mathcal{D} + 4b \sin^2 \frac{\pi j}{n} \right)$$

$$\Phi_n(1, 1)^2 = \prod_{\substack{1 \le j \le n \\ \gcd(j,n)=1}} 4 \sin^2 \frac{\pi j}{n} \ge 1$$

$$\Phi_n(-1, 1)^2 = \prod_{\substack{1 \le j \le n \\ \gcd(j,n)=1}} 4 \cos^2 \frac{\pi j}{n} \ge 1$$

Using the definition of R, we have

$$a - 4b \cos^2 \frac{\pi j}{n} > \sqrt{R} \left| \cos \frac{\pi j}{n} \right| \qquad \text{if } b > 0$$

$$\mathcal{D} + 4b \sin^2 \frac{\pi j}{n} > \sqrt{R} \left| \sin \frac{\pi j}{n} \right| \qquad \text{if } b < 0$$

$$|\Phi_n(\alpha, \beta)| > \sqrt{R}^{\frac{\varphi(n)}{2}}$$

Thus, using the fact that $R \ge 4$, we have the following result.

Theorem 6.17 *A sufficient condition that U_n has a primitive divisor is that*

$$2^{\frac{\varphi(n)}{2}} \ge n$$

Theorem 6.18 *If $n \ge 2 \cdot 10^9$, then*

$$\varphi(n) > \frac{n}{\log n}$$

Proof Since $\varphi(n) = n \prod_{p|n} \left(\frac{p-1}{p} \right)$

$$\frac{n}{\varphi(n)} = \prod_{p|n} \left(\frac{p}{p-1} \right)$$

$$\log \frac{n}{\varphi(n)} = \log \prod_{p|n} \left(\frac{p-1}{p} \right)^{-1}$$

$$= -\log \prod_{p|n} \left(\frac{p-1}{p} \right)$$

It is enough to prove that

$$\frac{n}{\varphi(n)} < \log n$$

$$\log \frac{n}{\varphi(n)} < \log \log n$$

$$-\log \prod_{p|n} \left(1 - \frac{1}{p} \right) < \log \log n$$

From the analysis in Hardy and Wright [HW60, Chap. XXII],

$$-\log \prod_{p|n} \left(\frac{p-1}{p} \right) < \sum_{p|n} \frac{1}{p} + \frac{1}{2}$$

$$< \sum_{p \leq \log n} \frac{1}{p} + \frac{1}{2 \log \log n}$$

Let $\pi(x)$ be the number of primes $\leq x$. Then using $\pi(x) < \frac{2x}{\log x}$ and *Abel summation formula*,

$$\sum_{p \leq x} \frac{1}{p} < \frac{2}{\log x} + 2 \log \log x - 2 \log \log 2$$

$$-\log \prod_{p|n} \frac{p-1}{p} < 2 \log \log \log n + \frac{3}{\log \log n} + \frac{1}{2} - 2 \log \log 2$$

$$< \log \log n$$

∎

Theorem 6.19 *If* $n \leq 2 \cdot 10^9$,

$$\varphi(n) > \frac{n}{6}$$

Proof First, see that the smallest n for which $p_1 \cdots p_n$ is at least $2 \cdot 10^9$ is $n = 9$. So if $n \leq 2 \cdot 10^9$, then $\omega(n) \leq 8$.

$$\varphi(n) = n \prod_{p|n} \frac{p-1}{p}$$

$$\geq n \prod_{2 \leq p \leq 19} \frac{p-1}{p}$$

$$> n\frac{1}{6}$$

∎

We can see that for $n \geq 75$,

$$n > 12\frac{\log n}{\log 2}$$

$$\frac{\varphi(n)}{2} \log n > \log n$$

$$\log_2 n < \frac{\varphi(n)}{2}$$

$$2^{\frac{\varphi(n)}{2}} > n$$

So, for $75 \leq n \leq 2 \cdot 10^9$, we have $2^{\varphi(n)/2} > n$. Checking $30 < n < 75$, we can manually verify that the inequality still holds. But it fails for $n = 30$ since $\varphi(30) = 8$ and $2^4 < 30$. Therefore, we have the following result.

Theorem 6.20 *If $n > 30$, then U_n always has at least one primitive divisor.*

We are left with $n < 30$.

$$\Phi_0 = 0$$
$$\Phi_1 = 1$$
$$\Phi_2 = 1$$
$$\Phi_3 = c - b$$
$$\Phi_4 = c - 2b$$
$$\Phi_5 = c^2 - 3bc + 3b^2$$
$$\Phi_6 = c - 3b$$
$$\Phi_8 = \Phi_4(\Phi_4 + 4b) + b^2$$
$$\Phi_{10} = \Phi_5 - 2b\Phi_4$$
$$\Phi_{12} = \Phi_4^2 - 3b$$

From our assumptions, $a, \mathcal{D} > 0$, $b \neq 0$ and $\gcd(b, c) = 1$. So, U_3, U_4, U_5 always have primitive divisors. U_6 has a primitive divisor relatively prime to c unless $c =$

Table 6.1 Possible exceptional indices

n	$\varphi(n)$	$4^{\frac{\varphi(n)}{2}}$	$8^{\frac{\varphi(n)}{2}}$	$12^{\frac{\varphi(n)}{2}}$	$16^{\frac{\varphi(n)}{2}}$
4	2	2	2.8	3.5	4
5	4	4			
6	2	2	2.8	3.5	4
8	4	4	8		
9	6	8			
10	4	4	8		
12	4	4	8	12	
14	6	8			
16	8	16			
18	6	8			
20	8	16			
24	8	16			
30	8	16			

$5, b = 1, \mathcal{D} = 1$ or $c = 1, b = -1, \mathcal{D} = 5$ or $c = 9, b = 2, \mathcal{D} = 1$. In the following Table 6.1, we list all possible exceptional indices where $4^{\frac{\varphi(n)}{2}} < n$. If $R^{\frac{\varphi(n)}{2}} > n$, then we do not list it and keep it empty since then n is not exceptional.

Note that for $n > 6$ and $R = 16$, all cells are empty. This proves Theorem 6.16 for this case.

Next, we prove that 12 is not exceptional if $R \in \{8, 12\}$. Using $\gcd(b, c) = 1$, we get that $\gcd(\Phi_3, b) = 1$, $\gcd(\Phi_4, b) = \gcd(\Phi_6, b) = 1$. Since $\Phi_4 = \Phi_3 - b = \Phi_6 + b$, $\gcd(\Phi_{12}, \Phi_6) = \gcd(\Phi_{12}, \Phi_3) = \gcd(2b^2, \Phi_3) = \gcd(2, \Phi_3)$. So this is either 1 or 2 and similarly $\gcd(\Phi_{12}, \Phi_4)$ is either 1 or 3. If 2 or 3 divides Φ_{12}, then we are done. Otherwise $6 \nmid \Phi_{12}$, then Φ_{12} has a primitive divisor at least 5 since $\Phi_{12} \geq 8$. The only case left is when $2 \mid \Phi_{12}$ or $3 \mid \Phi_{12}$ but they are not primitive divisors, then we know that $2 \| \Phi_{12}$ or $3 \| \Phi_{12}$. Thus, $\frac{\Phi_{12}}{6}$ is an integer relatively prime to Φ_4, Φ_3, Φ_6. So, Φ_{12} has a primitive divisor. We can similarly show that 8 is not an exceptional index if $R = 8$. We have the following result.

Theorem 6.21 *If $R = 8$ then neither 8 nor 12 is an exceptional index.*

Then only two cases are left to deal with when $R = 4$. $c = 1, b = -1, \mathcal{D} = 5$ or $c = 5, b = 1, \mathcal{D} = 1$. The first case evidently produces the Fibonacci numbers and we can directly compute the exceptional indices. We see that F_6, F_{12}, F_{18} are the only exceptional Fibonacci numbers. In the second case, $U_n = F_n$ when n is even. Since all exceptional indices are even, 6, 12, 18 are the exceptional indices. This completely proves Theorem 6.16.

Durst [Dur59, Dur61] continued the work of Ward and proved the following improvement. Durst argues that *It has long been known that the Lucas sequence generated by $z^2 - 3z + 2$ has six as its only exceptional index* and *the Lehmer sequence generated by the same polynomial has no exceptional indices.* So in Theorem 6.16,

18 *should be deleted* as an exceptional index since $F_{18} = 2^3 \cdot 17 \cdot 19$. He proves the following result.

Theorem 6.22 *For real Lehmer sequences, 6 and 12 are the only exceptional indices. Moreover, 12 is exceptional only for the sequences determined by*

$$(c, b) \in \{(1, -1), (5, 1)\}$$

6 is exceptional only when

$$(c, b) = (-3k + 2^{s+2}, -k + 2^s)$$

where $s \geq 1$, $2^{s+1} > k$ and k is positive and odd.

So, we also have the result that for a particular odd k, there are infinitely many exceptional indices.

6.5 Primitive Divisors of Complex Lehmer Sequences

We have already mentioned that it was not known until 1962 whether complex Lehmer numbers must have primitive divisors or not. Here, we show a result by Schinzel [Sch62b] that shows complex Lehmer numbers have primitive divisors as well. The aim of this section is to prove the following result. We do not show all the way that U_n has a primitive divisor for $n > 30$ because the original proofs by Schinzel and Postnikova [L P74], Stewart [Ste77], Voutier [Vou95, Vou96, Vou98] cannot be exactly categorized as elementary.

Theorem 6.23 *Let α, β be complex numbers such that β/α is not a root of unity. Then there is a constant $n_0 = n_0(\alpha, \beta)$ dependent on α, β such that U_n has a primitive divisor for all $n > n_0$. Furthermore, this constant n_0 can be effectively computed.*

Unfortunately, the result relies on *some deep theorem of* Gelfond [Gel60, Page 174], according to Schinzel.

Theorem 6.24 (Gelfond) *Let a, b be algebraic numbers such that $\log a / \log b$ is irrational and η be an arbitrary constant. Then the inequality*

$$|x_1 \log a + x_2 \log b| < e^{-\log^{2+\eta} x}, \quad \text{where } x = |x_1| + |x_2| > 0$$

has no solution for integers x_1, x_2 for

$$x > x_0 \left(a, b, \frac{\log a}{\log b}, \eta \right)$$

where x_0 is a constant that can be effectively computed.

This result is indeed very deep with many influential consequences. For example, Gelfond [Gel40] proves that if a, b, c are real numbers in a finite algebraic field K, none of which is 0 or ± 1 and at least one of which is not a unit, then the equation

$$a^x + b^y = c^z, \quad \text{where}$$
$$|x| + |y| + |z| = t$$

has no solution in integers x, y, z for $t > t_0(a, b, c)$ except when $a = 2^u, b = 2^v, c = 2^w$. Here, u, v, w are rational and t_0 can be effectively computed.

The same assumptions for our proof in the case of real Lehmer sequences apply here. See List 6.4 for a reminder. The proof of Theorem 6.23 relies on the next result.

Theorem 6.25 *Let α, β be complex numbers such that β/α is not a root of unity. Then for any $\epsilon > 0$ and $n > N(\alpha, \beta, \epsilon)$,*

$$U_n > |\alpha|^{n - \log^{2+\epsilon} n} \tag{6.2}$$

$$\Phi_n(\alpha, \beta) > |\alpha|^{\varphi(n) - 2^{\omega(n)} \log^{2+\epsilon} n} \tag{6.3}$$

where $\varphi(n)$ and $\omega(n)$ is the Euler's totient function and number of prime factors of n, respectively. The number $N(\alpha, \beta, \epsilon)$ is a constant that can be effectively computed.

Proof Set $a = \beta/\alpha, b = 1, \log b = 2i\pi$ in Theorem 6.24. Then for integers x_1, x_2, where $x_1 > x_0(\beta/\alpha, 1, (\log(\beta/\alpha))/2i\pi, \eta)$, we have

$$\left| x_1 \log \frac{\beta}{\alpha} + x_2 \cdot 2i\pi \right| \geq e^{-\log^{2+\eta} cx_1} \tag{6.4}$$

where $c = |\log(\beta/\alpha)|/(2\pi) + 2$. If $|\theta - 2\pi k| \geq d$ for all integer k and real θ and $0 \leq d \leq 3$,

$$|\cos\theta + i\sin\theta - 1| \geq \frac{d}{2}$$

For $x_1 > x_0$, (6.4) gives

$$\left| \left(\frac{\beta}{\alpha}\right)^x - 1 \right| \geq \frac{1}{2} e^{-\log^{2+\eta} cx_1} \tag{6.5}$$

From the definition of U_n, regardless of the parity of n,

$$U_n \geq \frac{\alpha^n - \beta^n}{\alpha^2 - \beta^2}$$

Using (6.4), we get

$$U_n \geq \frac{\alpha^n}{\alpha^2 - \beta^2} \left| \left(\frac{\beta}{\alpha}\right)^n - 1 \right|$$

$$\geq \frac{\alpha^n}{\alpha^2 - \beta^2} \frac{1}{2} e^{-\log^{2+\eta} cn}$$

Then using (6.5), we see that for a proper choice of η, (6.2) follows.

Note that $(\alpha - \beta)^2 = (\alpha + \beta)^2 - 4\alpha\beta$ is a non-zero integer, $\Phi_n(\alpha, \beta) = \prod_{d|n} U_{\frac{n}{d}}^{\mu(d)}$ and $U_n \leq 2|\alpha^n|$. So

$$|U_n| \leq \left| \frac{\alpha^n - \beta^n}{\alpha - \beta} \right|$$

$$\leq \frac{|2\alpha^n|}{|\alpha - \beta|}$$

$$\leq 2|\alpha^n|$$

Therefore

$$\Phi_n(\alpha, \beta) = \frac{\displaystyle\prod_{\substack{d|n, d>N \\ \mu(\frac{n}{d})=1}} U_d}{\displaystyle\prod_{\substack{d|n \\ \mu(\frac{n}{d})=-1}} U_d}$$

$$> \prod_{\substack{d|n \\ \mu(\frac{n}{d})=1}} \frac{|\alpha|^{d-\log^{2+\epsilon} d}}{2|\alpha^d|}$$

Now, β/α is not a root of unity. Hence $\alpha\beta \neq 1$ and $|\alpha|^2 + |\beta|^2 \geq 2\alpha\beta$ implies that $|\alpha| \geq \sqrt{2}$. Also, recall that from (1.7)

$$n = \sum_{d|n} \varphi(d)$$

Applying Möbius inversion to this equation gives

$$\varphi(n) = \sum_{d|n} \mu\left(\frac{n}{d}\right) d$$

Then we have the following.

$$\log \Phi_n(\alpha, \beta) > \left(\log \prod_{\substack{d|n, d>N \\ \mu(\frac{n}{d})=1}} |\alpha|^{d-\log^{2+\epsilon} d} \right) - \left(\log \prod_{\substack{d|n \\ \mu(\frac{n}{d})=1}} 2|\alpha|^d \right)$$

$$= \sum_{\substack{d|n, d>N \\ \mu(\frac{n}{d})=1}} \log |\alpha|^{d-\log^{2+\epsilon} d} - 2 \sum_{\substack{d|n \\ \mu(\frac{n}{d})=1}} \log |\alpha|^d$$

$$= \left(\sum_{\substack{d|n, d>N \\ \mu(\frac{n}{d})=1}} (d - \log^{2+\epsilon} d) \right) \log |\alpha| - 2 \left(\sum_{\substack{d|n \\ \mu(\frac{n}{d})=1}} d \right) \log |\alpha|$$

$$= \left(\sum_{d|n} \mu\left(\frac{n}{d}\right) d - \sum_{\substack{d|n \\ \mu(\frac{n}{d})=-1}} \log^{2+\epsilon} d - \sum_{\substack{d|n \\ d \le N}} d - 2 \sum_{\substack{d|n \\ \mu(\frac{n}{d})=-1}} 1 \right) \log |\alpha|$$

$$\ge \left(\sum_{d|n} \mu\left(\frac{n}{d}\right) d - \sum_{\substack{d|n \\ \mu(\frac{n}{d})=-1}} \log^{2+\eta} n - \sum_{\substack{d \le N}} d - 2 \sum_{\substack{d|n \\ \mu(\frac{n}{d})=-1}} 1 \right) \log |\alpha|$$

$$\frac{\log |\Phi_n(\alpha, \beta)|}{\log |\alpha|} \ge \varphi(n) - \frac{N(N+1)}{2} - 2^{\omega(n)-1} \log^{2+\eta} n - \omega(n)$$

Now, take n large enough so that

$$\log^2 n > \frac{N(N+1)}{2} + 1$$

Then we have

$$\log_{|\alpha|} |\Phi_n(\alpha, \beta)| > \varphi(n) - 2^{\omega(n)} \log^{2+\eta} n$$

This proves the inequality. ∎

We will now prove Theorem 6.23. We will use $\epsilon = 1$ in (6.3). Since $|\alpha| \ge \sqrt{2}$,

$$\Phi_n(\alpha, \beta) > \sqrt{2}^{\varphi(n)-2^{\omega(n)} \log^3 n}$$

We will be done if we can prove that

$$2^{\frac{\varphi(n)-2^{\omega(n)}\log^3 n}{2}} > n$$

$$\iff \varphi(n) - 2^{\omega(n)}\log^3 n > 2\log_2 n$$

$$= 2\frac{\log n}{\log 2}$$

From Theorem 6.18, for $n \geq 2 \cdot 10^9$,

$$\varphi(n) > \frac{n}{\log n}$$

Also, $2^{\omega(n)-1} \leq \sqrt{n}$ and

$$\frac{n}{\log n} - 2\sqrt{n}\log^3 n > \frac{2\log n}{\log 2}$$

holds for $n > 10^{20}$. Setting $n = \max(N, 10^{20})$ completes the proof.

References

[And14]   T. Andreescu, *Mathematical Reflections: The First Two Years* (XYZ Press, 2014)

[Bak72]   A. Baker, A sharpening of the bounds for linear forms in logarithms. Acta Arith. **21**, 117–129 (1972). https://doi.org/10.4064/aa-21-1-117-129

[Bak73]   A. Baker, A sharpening of the bounds for linear forms in logarithms II. Acta Arith. **24**(1), 33–36 (1973). https://doi.org/10.4064/aa-24-1-33-36

[Ban86]   A.S. Bang, TALTHEORETISKE UNDERSØGELSER. Tidsskr. Math. **4**, 70–80 (1886). ISSN: 09092528, http://www.jstor.org/stable/24539988

[Bil12]   M. Billal, Exponent GCD lemma. Math. Reflect. **6** (2012)

[BHV01]   Y. Bilu, G. Hanrot, P.M. Voutier, Existence of primitive divisors of Lucas and Lehmer numbers. J. Reine Angew. Math. (Crelles Journal) **2001**(539) (2001). https://doi.org/10.1515/crll.2001.080

[BV04]   G.D. Birkhoff, H.S. Vandiver, On the integral divisors of $a^n - b^n$. Ann. Mat. **5**(4), 173–180 (1904). ISSN: 0003486X. http://www.jstor.org/stable/2007263

[Car13]   R.D. Carmichael, On the numerical factors of the arithmetic forms $\alpha^n \pm \beta^n$. Ann. Mat. **15**(1/4), 30–48 (1913). ISSN: 0003486X. http://www.jstor.org/stable/1967797

[Dur59]   L.K. Durst, Exceptional real Lehmer sequences. Pacific J. Math. **9**(2), 437–441 (1959). https://projecteuclid.org:443/euclid.pjm/1103039266

[Dur61]   L.K. Durst, Exceptional real Lucas sequences. Pacific J. Math. **11**(2), 489–494 (1961). https://projecteuclid.org:443/euclid.pjm/1103037329

[Gel40]   A. Gelfond, Sur la Divisibilité de la Différence des Puissances de Deux Nombres Entiers par Une Puissance d'Un Idéal Premier. Rec. Math. [Mat. Sbornik] N.S. **7**(49)(1), 7–25 (1940). https://zbmath.org/?q=an:0023.10405|66.0183.05

[Gel60]   A.O. Gelfond, *Transcendental and Algebraic Numbers*. Transl. from the 1st Russian Ed. by Leo F. Boron (Dover, Mineola, 1960)

[HW60]   G.H. Hardy, E.M. Wright, *An Introduction to the Theory of Numbers* (Clarendon Press, Oxford, 1960)

[L P74]   A. Schinzel, L.P. Postnikova, Primitive divisors of the expression $A^n - B^n$ in algebraic number fields. J. Reine Angew. Math. (Crelles Journal) **1974**(268–269), 27–33 (1974). https://doi.org/10.1515/crll.1974.268-269.27

[Leh30]   D.H. Lehmer, An extended theory of Lucas' functions. Ann. Mat. **31**(3), 419–448 (1930).
           ISSN: 0003486X, http://www.jstor.org/stable/1968235
[Rot62]   A. Rotkiewicz, On Lucas numbers with two intrinsic divisors. Bull. Acad. Polon. Sci.
           Sér. Sci. Math. Astronom. Phys. **10**, 229–232 (1962)
[Sch62a]   A. Schinzel, On primitive prime factors of $a^n - b^n$. Math. Proc. Camb. Philos. Soc.
           58(4), 556–562 (1962). https://doi.org/10.1017/S0305004100040561
[Sch62b]   A. Schinzel, The intrinsic divisors of Lehmer numbers in the case of negative discrim-
           inant. Ark. Mat. **4**(5), 413–416 (1962). https://doi.org/10.1007/BF02591623
[Sch63a]   A. Schinzel, On primitive prime factors of Lehmer numbers I. Eng. Acta Arith. **8**(2),
           213–223 (1963). http://eudml.org/doc/207291
[Sch63b]   A. Schinzel, On primitive prime factors of Lehmer numbers II. Eng. Acta Arith. **8**(2),
           251–257 (1963). http://eudml.org/doc/207294
[Sch68]   A. Schinzel, On primitive prime factors of Lehmer numbers III. Eng. Acta Arith. **15**(1),
           49–70 (1968). http://eudml.org/doc/204877
[Ste77]   C.L. Stewart, Primitive divisors of Lucas and Lehmer numbers, in *Transcendence The-
           ory: Advances and Applications* (Cambridge, 1976) (1977), pp. 79–92
[Vou95]   P.M. Voutier, Primitive divisors of Lucas and Lehmer sequences. Math. Comput.
           64(210), 869–869 (1995). https://doi.org/10.1090/s0025-5718-1995-1284673-6
[Vou96]   P.M. Voutier, Primitive divisors of Lucas and Lehmer sequences, II. J. Théorie Nr. Bordx.
           8(2), 251–274 (1996). https://doi.org/10.5802/jtnb.168
[Vou98]   P.M. Voutier, Primitive divisors of Lucas and Lehmer sequences, III. Math. Proc. Camb.
           Philos. Soc. **123**(3), 407–419 (1998). https://doi.org/10.1017/s0305004197002223
[War55]   M. Ward, The intrinsic divisors of Lehmer numbers. Ann. Mat. **62**(2), 230–236 (1955).
           ISSN: 0003486X. http://www.jstor.org/stable/1969677
[Yab01]   M. Yabuta, A simple proof of Carmichael's theorem on primitive divisors. Fibonacci Q.
           39(5), 439–443 (2001). https://www.fq.math.ca/Scanned/39-5/yabuta.pdf
[Zsi92]   K. Zsigmondy, Zur Theorie der Potenzreste. Monatshefte Math. Phys. **3**(1), 265–284
           (1892). https://doi.org/10.1007/bf01692444

Chapter 7
Exercises

This book is not fully committed to Olympiad purposes. However, a lot of the theories in this topic can be used for Olympiad purposes and that has been the case for a while. So we will discuss some problems and see how to solve them. As we will see they are really helpful in solving problems regarding divisibility, Diophantine equations, etc. Especially, lifting the exponent lemma, Zsigmondy's theorem and cyclotomic polynomials are useful for solving many problems taken from contests around the world including the International Mathematical Olympiad (IMO) and the shortlists. Let us start with the problem we stated at the start of divisibility sequences, Chap. 3.

Problem 7.1 Prove that, for any positive integer n,

$$(x-1)(x^2-1)\cdots(x^{2014}-1) \mid (x^n-1)(x^{n+1}-1)\cdots(x^{n+2013}-1)$$

Solution We have already proven the generalized version of this in Theorem 3.13. We simply have to consider the cyclotomic polynomial.

$$x^n - 1 = \prod_{d\mid n} \Phi_d(x)$$

$$\prod_{i=1}^{n}(x^i-1) = \prod_{i=1}^{n}\prod_{d\mid i} \Phi_d(x)$$

$$= \prod_{i=1}^{n} \Phi_i(x)^{\left\lfloor \frac{n}{i} \right\rfloor}$$

Then we can simply write the quotient as follows:

$$\frac{(x^n - 1)(x^{n+1} - 1) \cdots (x^{n+2013} - 1)}{(x - 1)(x^2 - 1) \cdots (x^{2014} - 1)} = \frac{(x - 1)(x^2 - 1) \cdots (x^{n+2013} - 1)}{(x - 1)(x^2 - 1) \cdots (x^{2014} - 1)(x - 1)(x^2 - 1) \cdots (x^{n-1} - 1)}$$

$$= \frac{\prod_{i=1}^{n+2013}(x^i - 1)}{\prod_{i=1}^{2014}(x^i - 1) \prod_{i=1}^{n-1}(x^i - 1)}$$

$$= \frac{\prod_{i=1}^{n+2013} \Phi_i(x)^{\left\lfloor \frac{n}{i} \right\rfloor}}{\prod_{i=1}^{2014} \Phi_i(x)^{\left\lfloor \frac{n}{i} \right\rfloor} \prod_{i=1}^{n-1} \Phi_i(x)^{\left\lfloor \frac{n}{i} \right\rfloor}}$$

$$= \prod_{i=1}^{n+2013} \Phi_i(x)^{\left\lfloor \frac{n+2013}{i} \right\rfloor - \left\lfloor \frac{2014}{i} \right\rfloor - \left\lfloor \frac{n-1}{i} \right\rfloor}$$

Here, $\left\lfloor \dfrac{n+2013}{i} \right\rfloor - \left\lfloor \dfrac{2014}{i} \right\rfloor - \left\lfloor \dfrac{n-1}{i} \right\rfloor$ is a non-negative integer due to $\lfloor x + y \rfloor \geq \lfloor x \rfloor + \lfloor y \rfloor$.

Problem 7.2 A sequence is given by $a_1 = a_2 = 1$ and $a_n = a_{n-1} + ka_{n-2}$ for $n > 2$ where $k \neq 0$ is an integer. Show that $n \mid a_n$ for infinitely many n.

Solution The characteristic equation of (a) is $x^2 - x - k = 0$. It has discriminant $\Delta = 1 + 4k$. For a prime p, using $\epsilon = \left(\dfrac{\Delta}{p}\right)$ and $\sigma = \left(\dfrac{c}{p}\right)$, we obtain $\epsilon = \left(\dfrac{1 + 4k}{p}\right)$ and $\sigma = \left(\dfrac{1^2}{p}\right) = 1$. We can see that (a) is actually a Lucas with $a = 1, b = -k$. If n is a positive integer such that $\gcd(n, k) = 1$ and the canonical prime factorization of n is $n = \prod_{i=1}^{r} p_i^{e_i}$, then by Theorem 5.16,

$$n \mid a_{\psi(n)}$$

where

$$\psi(n) = \mathrm{lcm}\left(2, \prod_{i=1}^{r} p_i^{e_i - 1}\left(p_i - \left(\frac{c\Delta}{p_i}\right)\right)\right)$$

In fact, we do not even need the full factorization of n. We can simply consider a prime factor p of Δ so that $\epsilon = 0$. Then we have $p \mid a_p$ by Theorem 5.10. From Theorem 5.6, we have $p^i \mid a_{p^i}$ for all positive integer i. We have infinitely many such n for $n = p^k$ where p is a prime factor of Δ and k is an arbitrary positive integer.

Problem 7.3 Let $a_0 = 0, a_1 = 1$ and $a_n = ua_{n-1} - va_{n-2}$, where u, v are relatively prime integers. If p is an odd prime such that $p^2 \mid a_p$, show that $p = 3$.

Solution Fortunately, this follows directly from Theorem 5.13 once we notice that $a_n = U_n$, where (U) is a Lucas sequence of the first kind.

Problem 7.4 (*Vietnam 2020*) Let (a) be an integer sequence such that $a_1 = 5, a_2 = 13$, and $a_{n+1} = 5a_n - 6a_{n-1}$ for $n > 1$. Prove that

(a) $\gcd(a_n, a_{n+1}) = 1$ for all $n \in \mathbb{N}$.
(b) If p is a prime factor of a_{2^k}, then $2^{k+1} \mid p - 1$.

Solution First, we will show that $a_n = 2^n + 3^n$. We can prove it by induction or by using the characteristic polynomial of (a). $a_1 = 5, a_2 = 13$ so the base case holds. Assume that $a_m = 2^m + 3^m$ for $m < n$.

$$
\begin{aligned}
a_n &= 5a_{n-1} - 6a_{n-2} \\
&= 5(2^{n-1} + 3^{n-1}) - 6(2^{n-2} + 3^{n-2}) \\
&= 10 \cdot 2^{n-2} + 15 \cdot 3^{n-1} - 6 \cdot 2^{n-2} - 6 \cdot 3^{n-2} \\
&= 4 \cdot 2^{n-2} + 9 \cdot 3^{n-2} \\
&= 2^n + 3^n
\end{aligned}
$$

We can use Euclidean algorithm for proving $\gcd(a_n, a_{n+1}) = 1$. But first, we need to show that $\gcd(6, a_n) = 1$. This follows from the observation that neither 2 nor 3 divides a_n, since $2 \mid 2^n + 3^n$ would imply that $2 \mid 3^n$, for example.

$$
\begin{aligned}
\gcd(a_n, a_{n+1}) &= \gcd(a_n, 5a_n - 6a_{n-1}) \\
&= \gcd(a_n, 6a_{n-1}) \\
&= \gcd(a_n, a_{n-1}) \\
&= \cdots \\
&= \gcd(a_2, a_1) \\
&= \gcd(5, 13)
\end{aligned}
$$

Thus, $\gcd(a_n, a_{n+1}) = 1$. For the second part, assume that p is a prime divisor of a_n. Clearly, p is odd and p cannot be 2 or 3.

$$
\begin{aligned}
p &\mid 2^n + 3^n \\
2^n + 3^n &\equiv 0 \pmod{p} \\
2^n &\equiv -3^n \pmod{p} \\
2^{2n} &\equiv 3^{2n} \pmod{p}
\end{aligned}
$$

From Fermat's little theorem, we get that $2^{p-1} \equiv 1 \pmod{p}$ and $3^{p-1} \equiv 1 \pmod{p}$.

$$
\begin{aligned}
2^{p-1} &\equiv 3^{p-1} \pmod{p} \\
2^{\gcd(2n, p-1)} &\equiv 3^{\gcd(2n, p-1)} \pmod{p}
\end{aligned}
$$

Setting $n = 2^k$, we have that

$$2^{\gcd(2^{k+1},p-1)} \equiv 3^{\gcd(2^{k+1},p-1)} \pmod{p}$$

Let g be $\gcd(2^{k+1}, p-1)$. So g is of the form 2^d for some non-negative integer d. However, if $d < k+1$, then

$$2^{2^d} \equiv 3^{2^d} \pmod{p}$$

Multiplying the exponents by 2^{k-d} and raising both sides to 2^k, we get

$$2^{2^k} \equiv 3^{2^k} \pmod{p}$$
$$2^{2^k} \equiv -2^{2^k} \pmod{p}$$

since $p \mid 2^{2^k} + 3^{2^k}$. But this gives us $p \mid 2^{2^k+1}$, a contradiction. Therefore, d must be $k+1$ and $2^{k+1} \mid p-1$.

Problem 7.5 (*India 2001*) Let $a_1, a_2, \ldots$ be a strictly increasing sequence such that for any two positive integers m and n, $\gcd(a_m, a_n) = a_{\gcd(m,n)}$ holds. Let k be the least positive integer for which there exist positive integers $r < k$ and $s > k$ such that $a_k^2 = a_r a_s$. Prove that $r \mid k$ and $k \mid s$.

Solution Let d be $\gcd(r, s)$, g be $\gcd(k, r)$ and h be $\gcd(k, s)$. Note the following.

$$\gcd(a_k^2, a_r^2) = \gcd(a_k, a_r)^2$$
$$\gcd(a_r a_s, a_r^2) = a_{\gcd(k,r)}^2$$
$$a_r \gcd(a_s, a_r) = a_g^2$$
$$a_r a_d = a_g^2$$

If $g < r$, then $a_g < a_r$ since (a) is strictly increasing. We cannot have $g < d$ as well because then $a_g < a_d$ would imply $a_g^2 < a_r a_d = a_g^2$, which is impossible. Thus, $g < r$ and $g > d$ which gives us a smaller solution $g < k$, again impossible since k is the smallest such solution. Therefore, $g = r$ must hold and $r \mid k$. Using $\gcd(a_r, a_k) = a_{\gcd(r,k)}$, we get that $a_r \mid a_k$. Letting $a_k = a_r u$, $a_k a_r u = a_r a_s$ or $a_s = a_k u$. This shows that $a_k \mid a_s$ and

$$\gcd(a_k, a_s) = a_k$$
$$a_{\gcd(k,s)} = a_k$$

Thus, $k = \gcd(k, s)$ and $k \mid s$.

Problem 7.6 (*Hungary 2000*) Find all primes p and positive integers x, y such that

$$x^3 + y^3 = p^n$$

Solution Since $x + y \geq 2$, it has at least one prime divisor. If $x = y$, then $p = 2$ and $x^3 = 2^{n-1}$. We have the solution $x = 2^k, n = 3k + 1$. Next, we assume that $x \neq y$. Let g be $\gcd(x, y)$. Since g^3 divides p^n, g must be a power of p. Letting $g = p^u, x = ga, y = gb$ where $\gcd(a, b) = 1$, we get

$$p^{3u}(a^3 + b^3) = p^n$$
$$a^3 + b^3 = p^{n-3u}$$
$$(a + b)(a^2 - ab + b^2) = p^{n-3u}$$

Here, $a + b$ must be a power of p as well. If $(a, b) \notin \{(2, 1), (1, 2)\}$, then by Zsigmondy's theorem $a^3 + b^3$ has a prime divisor which does not divide $a + b$. This leads to a contradiction since both $a + b$ and $a^3 + b^3$ are powers of p. So we must have that $(a, b) \in \{(1, 2), (2, 1)\}$ and $p = 3, n - 3u = 2$.

Problem 7.7 (*Ha Noi Team Selection Test for VMO 2019*) Find all natural numbers m, n such that

$$5^m = n \cdot 2^m + 1$$

Solution We can rewrite the problem as $2^m \mid 5^m - 1$ and n as the quotient. If $m = 1$, then $n = 2$ is a solution. Let us assume $m > 1$. Since $4 \mid 5 - 1$, by Theorem 6.2,

$$v_2(5^m - 1) = v_2(5 - 1) + v_2(m)$$
$$v_2(2^m) = v_2(4) + v_2(m)$$
$$m = 2 + v_2(m)$$

Let us write $m = 2^k s$ where s is odd and $k \geq 0$ is an integer. Then the equation above gives

$$2^k s = 2 + k$$

If $k \geq 1$, then k must be even. Letting $k = 2l$ where $l \geq 1$ is an integer, we have

$$2^{2l} s = 2 + 2l$$
$$2^{2l-1} s = l + 1$$

Here, $l \geq$ so l must be odd since $l + 1$ is divisible by 2. The only solution to this equation is $(l, s) = (1, 1)$ since

$$2^{2l-1} s \geq 2^{2l-1}$$
$$> l + 1$$

for $l \geq 3$. Thus, we have two valid solutions $(k, s) = (0, 2), (2, 1)$ in this case which gives $(m, n) \in \{(1, 2), (2, 6), (4, 39)\}$.

We leave the similar problem below as an exercise.

Problem 7.8 (*Croatia Team Selection Test 2006*) Find all positive integer n such that

$$2^n \mid 3^n - 1$$

Problem 7.9 For an odd prime p, prove that

$$(p - 1)!^{p^{n-1}} \equiv -1 \pmod{p^n}$$

Solution By Wilson's theorem, we have

$$(p - 1)! \equiv -1 \pmod{p}$$

So $p \mid (p - 1)! + 1$. Since p is an odd prime, by Theorem 6.2,

$$\nu_p((p - 1)!^{p^{n-1}} + 1) = \nu_p((p - 1)! + 1) + \nu_p(p^{n-1})$$
$$\geq 1 + n - 1 = n$$

Thus, $p^n \mid (p - 1)^{p^{n-1}} + 1$.

Problem 7.10 When does the equation $p^a - 1 = 2^n(p - 1)$ have a solution in positive integers for a prime p?

Solution First, p cannot be 2 since $p^a - 1$ is odd and $2^n(p - 1)$ is even. Clearly a cannot be 1 since

$$p - 1 = 2^n(p - 1)$$
$$\geq 2(p - 1)$$
$$> p - 1$$

The last inequality is true because $p > 2$. If $a = 2$, then

$$p^2 - 1 = 2^n(p - 1)$$
$$p + 1 = 2^n$$

So for every positive integer n for which p is a prime is a solution to the equation. If $a > 2$, then by Theorem 6.5, $p^a - 1$ has a divisor q which does not divide $p - 1$. Since q is odd, it does not divide $2^n(p - 1)$ either. This would lead to a contradiction, so the only solution is available when $a = 2$ and $p = 2^n - 1$ is a prime.

Remark The prime numbers of the form $M_n = 2^n - 1$ are called *Mersenne prime*. It is not known yet exactly which values of n produce prime M_n.

Problem 7.11 Find all positive integer solutions to the equation

$$(a+1)(a^2 + a + 1) \cdots (a^n + a^{n-1} + \cdots + 1) = a^m + a^{m-1} + \cdots + 1$$

Solution We can rewrite the equation as follows.

$$\frac{a^2 - 1}{a - 1} \frac{a^3 - 1}{a - 1} \cdots \frac{a^{n+1} - 1}{a - 1} = \frac{a^{m+1} - 1}{a - 1}$$

$$(a^2 - 1)(a^3 - 1) \cdots (a^{n+1} - 1) = (a^{m+1} - 1)(a - 1)^{n-1}$$

From the first equation, it is clear that if $m = n$ or $n = 1$, then $m = n = 1$, and also that $m < n$ cannot hold. So $m > n > 1$, which gives $m \geq 3$. Then $a^{m+1} - 1$ has a primitive divisor p that does not divide any of $a^2 - 1, a^3 - 1, \ldots, a^{n+1} - 1$ unless $a = 2, m + 1 = 6$.

$$(2^2 - 1)(2^3 - 1) \cdots (2^{n+1} - 1) = 2^6 - 1$$

Here, $n < 4$ since $2^4 - 1$ is divisible by 5 which does not divide $2^6 - 1$. For $n = 3$, the left side is divisible by 3 only whereas the right side is divisible by 3^2, again impossible.

Problem 7.12 Let (a) be a divisibility sequence. Show that (a) is a strong divisibility sequence if and only if for any prime p and positive integers m, n such that $p \mid a_m$ and $p \mid a_n$, we have $p \mid a_{\gcd(m,n)}$.

Solution Recall that if (a) is a divisibility sequence, then by Theorem 3.3, (a) is a strong divisibility sequence if and only if for any prime p and positive integers α, k, we have $p^\alpha \mid a_k$ if and only if $\rho_\alpha \mid k$. A consequence of this theorem is that every prime p has a unique rank of apparition in (a). In other words, (a) is a strong divisibility sequence if and only if the rank of apparition of every prime p is unique.

To prove the backward implication, let k be the least index with $p \mid a_k$. Then k is the unique rank of apparition of p in (a), since $p \mid a_n$ implies

$$p \mid \gcd(a_k, a_n) \mid a_g$$

where $g = \gcd(k, n) \leq k$. Thus $g = k$ and $k \mid n$ for any such n. This fulfills the requirement for (a) to be a strong divisibility sequence. For the other direction, suppose that (a) is a strong divisibility sequence and ρ is the rank of apparition of p. If $p \mid a_m$ and $p \mid a_n$, then $\rho \mid m$ and $\rho \mid n$. Hence $\rho \mid \gcd(m, n)$, whence $p \mid a_{\gcd(m,n)}$.

Problem 7.13 Find all n such that V_n does not have a primitive divisor, where (V) is a Lucas sequence of the second kind.

Solution Note that $U_{2n} = U_n V_n$ and $\gcd(U_n, V_n) \in \{1, 2\}$. So V_n has a primitive divisor if and only if U_{2n} has a primitive divisor. U_{2n} has a primitive divisor except when $n \in \{1, 2, 6\}$ or $n = 12, a = 1, b = -1$.

The next problem from IMO 1990 is very instructional.

Problem 7.14 (*IMO 1990, p. 234 in Dušan, Vladimir, and Matic* [DVM11]) Find all positive integers n such that

$$n^2 \mid 2^n + 1$$

Solution We will try to understand what prime factors n is comprised of. Clearly, n is odd and so any prime factor of n is odd as well. Let p be the smallest prime factor of n. Also, $n = 1$ is a trivial solution so we can assume that $n > 1$. Then using $n^2 \mid 2^n + 1$, we get

$$p^2 \mid 2^n + 1$$
$$2^n + 1 \equiv 0 \pmod{p}$$
$$2^{2n} \equiv 1 \pmod{p}$$

By Fermat's little theorem, $2^{p-1} \equiv 1 \pmod{p}$. So

$$2^g \equiv 1 \pmod{p}$$

where $g = \gcd(p - 1, 2n)$. If $g > 2$, then g must have a prime factor less than p. This would contradict the minimality of p. Thus $g = 2$ and $2^2 \equiv 1 \pmod 3$, which implies $p = 3$. Before we move on to the next prime factor, we have to calculate exactly to which power 3 divides n. We can use lifting the exponent lemma for this. If $3^\alpha \parallel n$, then

$$\nu_3(2^n + 1) \geq \nu_3(n^2)$$
$$\nu_3(2 + 1) + \nu_3(n) \geq \nu_3(3^{2\alpha})$$
$$1 + \alpha \geq 2\alpha$$

This gives $\alpha = 1$. Next, assume that $n = 3m$ where $3 \nmid m$. Let q be the smallest prime factor of m. In a similar fashion as before,

$$q^2 \mid 2^{3m} + 1$$
$$8^m + 1 \equiv 0 \pmod{q^2}$$
$$8^{\gcd(2m, q-1)} \equiv 1 \pmod{q}$$

Here, we again have that $\gcd(2m, q - 1) = 2$ by the same reasoning. Therefore, q divides $8^2 - 1 = 3^2 \cdot 7$. Since q is different from p, only possible value of q is 7. But this is impossible as it would imply

$$8^m + 1 \equiv 0 \pmod 7$$
$$1^m + 1 \equiv 0 \pmod 7$$
$$2 \equiv 0 \pmod 7$$

Thus, q cannot have any prime factors and the only possible value of q is 1. Thus $n = 3$ is the only solution other than 1.

Problem 7.15 (*IMO 2000 Shortlist, Problem N4, p. 302 in Dušan, Vladimir, and Matic* [DVM11]) Find all positive integers a, m, n such that

$$a^m + 1 \mid (a + 1)^n$$

Solution From the divisibility relation, we have that $a^m + 1$ can only have prime divisors which also divide $a + 1$. Let p be such a prime divisor.

$$a + 1 \equiv 0 \pmod p$$
$$a^m + 1 \equiv 0 \pmod p$$
$$(-1)^m + 1 \equiv 0 \pmod p$$

First, assume that $a + 1$ has at least one odd prime divisor. Let p be such a prime. Then m has to be odd since otherwise $(-1)^m + 1 \equiv 2 \pmod p$ would give us a contradiction. If $a \neq 1$ or $(a, m) \neq (2, 3)$, then by Zsigmondy's theorem, $a^m + 1$ has a prime divisor that does not divide $a + 1$. This would again lead to a contradiction, so this cannot happen either. We have $(a, m) \in \{(1, m), (2, 3)\}$. If $a = 1, m = 1$, then any n satisfies the relation. If $a = 2, m = 3$, then any $n \geq 2$ works.

Now, $a + 1$ has no odd prime divisor, so $a + 1 = 2^k$ for some positive integer k. If m is even, then

$$a^m + 1 \equiv 2 \pmod 4$$

So $a^m + 1$ has an odd factor other than 2 if $a^m + 1 > 2$. But $a + 1$ only has 2 as prime factor, so this cannot hold true. We are left with odd m and $a + 1 = 2^k$. Since $a > 1$, we have $k > 1$ as well. In this case, $a^m + 1$ again has a prime factor that does not divide $a + 1 = 2^k$ by Zsigmondy's theorem since $a > 1$ and odd. Therefore, this is impossible and we have found all the solutions above.

Another relevant problem appeared in the shortlist of IMO 2000 which we can discuss here, see Dušan, Vladimir, and Matic [DVM11, p. 302, Problem $N4$].

Problem 7.16 Does there exist a positive integer n such that $n \mid 2^n + 1$ and has exactly 2000 distinct prime divisors?

Solution We can see that $n = 3$ satisfies $n \mid 2^n + 1$ and has exactly 1 prime divisor. Now, $3^2 \parallel 2^3 + 1$ and by Theorem 6.2,

$$\nu_3\left(2^{3^{k+1}}+1\right) = \nu_3\left(\left(2^3\right)^{3^k}+1\right)$$
$$= \nu_3\left(2^3+1\right)+\nu_3(3^k)$$
$$= 2+k$$

Call n a k-good number if $n \mid 2^n + 1$ and has exactly k distinct prime divisors. By Theorem 6.5, for every positive integer k, $2^{3^{k+1}} + 1$ has a primitive prime divisor that does not divide $2^{3^k} + 1$. Let p_{k+1} be such a prime divisor. Then $p_1 = 3$ and $p_{k+1} \notin \{p_1, p_2, \ldots, p_k\}$ for every k and p_i divides $2^{3^{k+1}} + 1$ for $1 \le i \le k+1$. Setting $k \to 1999$ and

$$n = 3^{2000} p_2 \cdots p_{2000}$$

we can see that n is 2000-good.

Remark The problem would be a lot easier if it asked for only 2000 prime divisors (not necessarily distinct). We would only have to prove that $3^k \mid 2^{3^k} + 1$.

The next couple of problems can be solved using a similar idea.

Problem 7.17 (*Vietnam Team Selection Test 2020*) Find all positive integer k such that there are only finitely many positive odd numbers n for which $n \mid k^n + 1$.

Problem 7.18 Show that there are infinitely many positive integers n for which $n \mid 2^n + 2$.

Problem 7.19 Let a, b, m, n be positive integers. Prove that there exists a prime p such that p divides both $a^m - b^m$ and $a^n - b^n$ if and only if $m \mid n$ or $n \mid m$.

Solution Let $g = \gcd(m, n)$. Since p divides both $a^m - b^m$ and $a^n - b^n$, p also divides $a^g - b^g$. If neither $m \mid n$ nor $n \mid m$ holds, then $g < \min(m, n)$. However, then by Theorem 6.5, $a^m - b^m$ or $a^n - b^n$ would have a prime divisor q which does not divide $a^g - b^g$. So in order for $m \mid n$ or $n \mid m$ to hold, we only have to consider such a primitive divisor. Take such a primitive prime divisor and call it p. This p is the desired prime in the statement.

Problem 7.20 Let a be a fixed positive integer. Prove that $\dfrac{p-1}{\mathrm{ord}_p(a)}$ is unbounded where p runs through all the primes that does not divide a.

Solution Let p be a fixed prime. Consider a prime divisor q of $\Phi_p(a)$. By Theorem 1.45, either $q \mid p$ or $q \equiv 1 \pmod{p}$. If $d = \mathrm{ord}_q(a)$, then $d \mid p$ so $d \le p$. Now, using Chinese Remainder Theorem and Theorem 1.47, we can see that for any positive integer k, there is a prime p_0 such that $p_0 + 1, 2p_0 + 1, \ldots, kp_0 + 1$ are all composite. Let p be such a prime p_0 which we fixed earlier. Then for such p, the minimum possible value of i for which $pi + 1$ can be the prime q is greater than k. Then $q > pk + 1$ and we have the following.

$$\frac{q-1}{\mathrm{ord}_q(a)} = \frac{q-1}{d}$$

$$\geq \frac{q-1}{p}$$

$$> \frac{pk}{k} = k$$

Thus, this ratio can be arbitrarily large since k can be chosen freely.

Problem 7.21 Let a, n be positive integers and p be a prime. If

$$\frac{a^n - 1}{a - 1} = p^k$$

for some positive integer k, then prove that n is the power of a prime.

Solution This problem is very easily solved with the help of cyclotomic polynomials. Note that we can rewrite the equation as the following.

$$\frac{\Phi_n(a)}{\Phi_1(a)} = p^k$$

$$\prod_{\substack{d|n \\ d>1}} \Phi_d(a) = p^k$$

For the sake of contradiction, assume that n is not a prime and $n = q^r m$ where $q \nmid m$. Both $\Phi_{qm}(a)$ and $\Phi_m(a)$ are powers of p, so by Theorem 1.48, $\frac{qm}{m}$ is a power of p. So, we must have $q = p$. Now, if $m > 1$, then $\Phi_q(a)$ and $\Phi_m(a)$ are powers of p as well. Again, Theorem 1.48 implies that $\frac{p}{m}$ or $\frac{m}{q}$ is a power of p, which is impossible since $p \nmid m$. Thus, $n = p^r$ for some positive integer r.

The next problem from Russia can be solved similarly.

Problem 7.22 (*Russian Olympiad 1996*) Let p be a prime and x, y, k be positive integers such that

$$x^n + y^n = p^k$$

Then prove that n is a power of prime.

Problem 7.23 Let $a > 2$ be a positive integer. Prove that $a^{a-1} - 1$ is never square-free.

Solution Let p be a prime divisor of $a - 1$.

$$\nu_p(a^{a-1} - 1) = \nu_p(a - 1) + \nu_p(a - 1)$$

$$= 2\nu_p(a - 1)$$

This is at least 2, so p^2 divides $a^{a-1} - 1$.

Problem 7.24 Let a, b, c be positive integers such that $a^n + b^n \mid c^n$ for all positive integer n. Prove that $a = b$.

Solution Let g be $\gcd(a, b)$ and $a = ax, b = gy$ with $\gcd(x, y) = 1$. The relation then converts into $g^n(x^n + y^n) \mid c^n$. We have that $g \mid c$. Letting $c = gd$, we have $x^n + y^n \mid d^n$ where $\gcd(x, y) = 1$. Consider an odd positive integers $n > 3$. If $x \neq y$, by Zsigmondy's theorem, $x^n + y^n$ has a prime divisor that does not divide any of $x + y, x^3 + y^3, \ldots, x^{n-2} + y^{n-2}$. So there are infinitely many primes p which divide d^n. This is impossible since $\mathrm{rad}(d^n) = \mathrm{rad}(d)$. So, we must have that $x = y$ and $a = b$.

Problem 7.25 (*JBMO 2017 Shortlist*) Find all positive integers n such that there exists a prime p for which $p^n - (p - 1)^n$ is a power of 3.

Solution Clearly, n cannot be 1. If $n = 2$, then $p = 2$ works. Now, $p = 3$ cannot be a solution for any positive integer n since that would imply that 3 divides $(p - 1)^n = 2^n$. For a prime $p > 3$, we can write p in the form $6k \pm 1$. If $p = 6k + 1$,

$$p^n - (p - 1)^n \equiv 1 \pmod 3$$

so it cannot be a power of 3. So p has to be of the form $6k - 1$.

$$p^n - (p - 1)^n \equiv 0 \pmod{3^k}$$
$$p^n \equiv (p - 1)^n \pmod{3^k}$$
$$\left(\frac{p}{p - 1}\right)^n \equiv 1 \pmod{3^k}$$

By Euler's theorem,

$$\left(\frac{p}{p - 1}\right)^{3^{k-1} \cdot 2} \equiv 1 \pmod{3^k}$$
$$\left(\frac{p}{p - 1}\right)^{\gcd(2 \cdot 3^{k-1}, n)} \equiv 1 \pmod{3^k}$$

Let g be $\gcd(2 \cdot 3^{k-1}, n)$. So g is of the form 3^u or $2 \cdot 3^u$ for some non-negative integer u. If $g = 3^u$ for some u then $u > 0$. Then $n = 3^u m$ for some positive integers m and u.

$$p^{3^u m} - (p - 1)^{3^u m} = 3^k$$
$$\left(p^{3^{u-1} m}\right)^3 - \left((p - 1)^{3^{u-1} m}\right)^3 = 3^k$$

By Theorem 6.5, $\left(p^{3^{u-1}m}\right)^3 - \left((p-1)^{3^{u-1}m}\right)^3$ has at least two distinct prime divisors since none of the exceptions are applicable here. We can solve the case $g = 2 \cdot 3^u$ in a similar manner.

Problem 7.26 Find all triplets (x, y, p) such that p is a prime and

$$2^x + p^y = 19^x$$

Solution Rewrite the equation as

$$p^y = 19^x - 2^x$$
$$= (19 - 2)(19^{x-1} + \cdots + 2^{x-1})$$

Clearly, 17 divides p^y, so $p = 17$. If $x > 1$, by Theorem 6.5, $19^x - 2^x$ has at least two prime divisors, a contradiction. So $x = 1$ and $y = 1$, $p = 17$ is the only solution.

We leave the following Diophantine equations as exercise.

Problem 7.27 (*JBMO 2013*) Find all positive integers x, y, z such that

$$20^x + 13^y = 2013^z$$

Problem 7.28 Find all positive integers x, y, z such that $3^x + 11^y = z^2$.

Problem 7.29 Solve the Diophantine equation

$$1997^a + 15^b = 2012^c$$

In 2011 (or maybe 2012), the first author posed the next problem to some of the IMO contestants (probably Dhananjoy Biswas and one or two other contestants) of Bangladesh at the IMO camp. The problem has also been part of the Olympiad marathon in the Art of Problem Solving (AoPS) forum.

Problem 7.30 Let a, n and d be positive integers such that $a \geq 3, n \geq 3, d > 1$ and $a, a + d, \ldots, a + (n-1)d$ are all primes. Prove that

$$\tau\left(2^{\lfloor \frac{d}{2} \rfloor} + 1\right) \geq 2^{2^{\pi(n)-1}-1}$$

where $\pi(x)$ is the number of primes $\leq x$ and $\tau(n)$ is the number of positive divisors of n.

Solution First of all, $a < n$ is impossible since otherwise $a + ad$ is divisible by a (recall that $a > 1$). So we have $a > n$. We will first show that d must be divisible by p for any prime $p \leq n$.

For the sake of argument, assume that $p \nmid d$. Since $p \le n$ and a is a prime, $p \nmid a$. Therefore, both a and d are relatively prime to p. Let e be the modular inverse of d modulo p such that $de \equiv 1 \pmod{p}$.

$$a + id \equiv ade + id \pmod{p}$$
$$\equiv d(ae + i) \pmod{p}$$

Here, $ae \pmod{p}$ is less than p but i runs through all of $0, \dots, n-1$. So, there is at least one i such that $p \mid ae + i$. Since $d > 1$ and $a + id$ is divisible by dp, we have that $a + id$ is not a prime, a contradiction. Thus, $p \mid d$ for every $p \le n$.

So, we can write $d = p_1 p_2 \cdots p_k L$ where $p_1 = 2$, $p_2 = 3$, $\dots$, p_k is the greatest prime $\le n - 1$ and L is a positive integer.

$$2^{\lfloor \frac{d}{2} \rfloor} + 1 = 2^{p_2 \cdots p_k L} + 1$$
$$= \left(2^L\right)^{p_2 \cdots p_k} + 1$$
$$= X^{p_2 \cdots p_k} + 1$$

The number of distinct prime divisors of $\lfloor \frac{d}{2} \rfloor$ is $k - 1$ where $k = \pi(n)$. Let $P = p_2 \cdots p_k$ and h, l be two distinct divisors of P. Since p_i is odd for $1 < i \le k$,

$$\gcd(X^h + 1, X^l + 1) = X^{\gcd(h,l)} + 1$$

By Zsigmondy's theorem, $X^h + 1$ and $X^l + 1$ both have primitive divisors that does not divide $X^{\gcd(h,l)} + 1$. If ρ, γ are the respective prime divisors, then we can see that both ρ and γ are different than any prime divisor of $X^{\gcd(h,l)} + 1$ and $\rho \ne \gamma$. Therefore, for a proper divisor h of P and a prime $q \mid P$, $X^{hq} + 1$ has a primitive divisor that does not divide $X^h + 1$ unless $X = 2, h = 1, q = 3$. In other words, we get a primitive divisor of $X^h + 1$ for every divisor h of P except when $h = 3$ (possibly). Then the number of distinct prime divisors of $X^{p_2 \cdots p_k} + 1$ is at least $\tau(P) - 1$. Since $P = p_2 \cdots p_k$ has $k - 1$ distinct prime divisors and $X^P + 1$ has at least $\tau(P) - 1$ distinct prime divisors,

$$\tau(P) = 2^{\tau(n)-1}$$
$$\tau\left(2^{\lfloor \frac{d}{2} \rfloor} + 1\right) \ge 2^{2^{\pi(n)-1} - 1}$$

In a similar manner, the following problems can be solved.

Problem 7.31 (*IMO 2002 shortlist, N3*) Let $p_1, \dots, p_n$ be primes greater than 3. Prove that $2^{p_1 \cdots p_n} + 1$ has at least 4^n divisors.

Remark From our solution to Problem 7.30, it is clear that this is a very weakened statement. A stronger statement would be: $2^{p_1 \cdots p_n} + 1$ has at least 2^{2^n} divisors where $p_1, \dots, p_n$ are distinct primes at least 5.

Problem 7.32 (*USAMO 2007*) Prove that $7^{7^n} + 1$ has at least $2n + 3$ prime (not necessarily distinct) divisors.

Remark This can be generalized: for any prime $p \equiv 3 \pmod 4$, $p^{p^n} + 1$ has at least $2n + 2$ prime (not necessarily distinct) divisors.

Problem 7.33 Find all pairs of positive integer (a, b) such that

$$a^b - 1 \mid b^a$$

Solution Clearly, a cannot be 1 and $(a, b) = (2, 1)$ is a solution. Let $a > 2, b > 1$ and p be a prime divisor of $a - 1$. First, consider the case p is odd. By Theorem 6.2,

$$\nu_p(a^b - 1) = \nu_p(a - 1) + \nu_p(b)$$

Since $p \mid a - 1$ and $a - 1 \mid b^a$, we also have $p \mid b$. Assume that $a - 1 = p^r c$ and $b = p^s d$ where $p \nmid c, d$.

$$\nu_p(b^a) \geq \nu_p(a^b - 1)$$
$$a\nu_p(b) \geq \nu_p(a - 1) + \nu_p(b)$$
$$(p^r c + 1)s \geq r + s$$

This holds for any odd prime p and positive integers r, s, c.

If $p = 2$ and $a \equiv 1 \pmod 4$,

$$\nu_2(a^b - 1) = \nu_2(a - 1) + \nu_2(b)$$
$$\nu_2(b^a) \geq \nu_2(a^b - 1)$$
$$a\nu_2(b) \geq \nu_2(a - 1) + \nu_2(b)$$

Letting $a - 1 = 2^k x$, we see that $a\nu_2(b) \geq \nu_2(a - 1) + \nu_2(b)$ holds for all b since

$$(2^k x + 1)\nu_2(b) \geq k + \nu_2(b)$$

is clearly true. Otherwise, if $b = 2^s d, 2 \parallel a - 1, a + 1 = 2^r c$ we get

$$\nu_2(a^b - 1) = \nu_2(a + 1) + \nu_2(a - 1) + \nu_2(b) - 1$$
$$\nu_2(b^a) \geq \nu_2(a^b - 1)$$
$$a\nu_2(b) \geq \nu_2(a + 1) + \nu_2(a - 1) + \nu_2(b) - 1$$
$$(2^r c - 1)s \geq r + 1 + s - 1$$
$$= r + s$$

Since $r \geq 2$, we have that the last inequality if true for r and s.

So, as long as we have $\mathrm{rad}(a - 1) = \mathrm{rad}(b)$, such a, b satisfy the condition.

Problem 7.34 Find all pairs of positive integer (a, b) such that

$$b^a \mid a^b - 1$$

Solution Clearly $a = 1$ is a solution for any positive integer b. Let p be the smallest prime factor of b. Since $p \mid a^b - 1$, we have $p \nmid a$.

$$a^b \equiv 1 \pmod{p}$$
$$a^{p-1} \equiv 1 \pmod{p}$$
$$a^{\gcd(b, p-1)} \equiv 1 \pmod{p}$$

Since b cannot have a prime factor smaller than p, it follows that $\gcd(b, p - 1) = 1$. Then $p \mid a - 1$. Let $a - 1 = p^u x$, where $p \nmid x$ and $b = p^v y$. If p is odd,

$$\nu_p(a^b - 1) \geq \nu_p(b^a)$$
$$\nu_p(a - 1) + \nu_p(b) \geq a\nu_p(b)$$
$$u + v \geq (p^u x + 1)v$$
$$u \geq p^u x v$$

This is impossible, so b or $a - 1$ cannot have odd prime factors, whence $a - 1 = 2^u, b = 2^v$. If $u \geq 2$,

$$\nu_2(a^b - 1) = \nu_2(a - 1) + \nu_2(b)$$
$$u + v \geq a\nu_2(b)$$
$$= (2^u + 1)v$$

This is again impossible. So $u = 1$ and $a = 3$.

$$\nu_2(3^b - 1) = \nu_2(3 - 1) + \nu_2(3 + 1) + \nu_2(b) - 1$$
$$1 + 2 + v - 1 \geq 3v$$
$$2 \geq 2v$$

So, $v = 1$ and $b = 2$.

Problem 7.35 Solve in positive integers:

$$2 \cdot 3^x = 5^y + 1$$

Solution If y is even,

$$5^y + 1 \equiv (-1)^{2b} + 1 \pmod{3}$$
$$2 \cdot 3^x \equiv 2 \pmod{3}$$

This is impossible, so y is odd. But by Theorem 6.5, $5^y + 1$ has a prime divisor which does not divide $5 + 1 = 2 \cdot 3$ if $y > 1$. So, $y = 1, x = 1$ is the only solution.

We leave the problems below as exercises.

Problem 7.36 (*Romania Team Selection Test 1994*) Let n be an odd positive integer. Prove that

$$\left((n-1)^n + 1\right)^2 \mid n(n-1)^{(n-1)^n + 1} + n$$

Problem 7.37 (*Romania TST 2007*) Prove that there exist infinitely many primes p and q such that $p \mid 2^{q-1} - 1$ and $q \mid 2^{p-1} - 1$.

Problem 7.38 Find all positive integers x, y and primes p such that

$$p^x - a^p = 1$$

Problem 7.39 (*Kazakhstan National Olympiad*) Find all triplets (k, m, n) such that

$$k^m \mid m^n - 1$$
$$k^n \mid n^m - 1$$

Reference

[DVM11] D. Djukić, V. Janković, I. Matic, *The IMO Compendium: A Collection of Problems Suggested for the International Mathematical Olympiads: 1959–2009* (Springer, 2011)

Glossary

GCD and LCM $\gcd(a, b)$ and $\mathrm{lcm}(a, b)$ are respectively the greatest common divisor and the least common multiple of a and b. $\gcd(a, \gcd(b, c)) = \gcd(\gcd(a, b), c)$ and $\mathrm{lcm}(a, \mathrm{lcm}(b, c)) = \mathrm{lcm}(\mathrm{lcm}(a, b), c)$. If $(a, b) = 1$ then $\gcd(a, bc) = \gcd(a, c)$. Also, if a positive integer d divides a and b, then $d \mid \gcd(a, b)$. $\gcd(ab, ac) = a \gcd(b, c)$ and $\mathrm{lcm}(ab, ac) = a \mathrm{lcm}(b, c)$. See Billal and Hossein [BH18, Chap. 1] for more on gcd and lcm.

Floor and Ceiling Functions $\lfloor x \rfloor$ and $\lceil x \rceil$ are respectively the floor and ceiling functions of x. $\lfloor x \rfloor$ is the greatest integer less than or equal to x. $\lceil x \rceil$ is the least integer greater than than or equal to x. $\lfloor \frac{n}{a} \rfloor = \lfloor \frac{n-1}{a} \rfloor$ if and only if $a \nmid n$. Note that

$$\lfloor x \rfloor + \lfloor y \rfloor + 1 \geq \lfloor x + y \rfloor \geq \lfloor x \rfloor + \lfloor y \rfloor$$

and

$$\lceil x \rceil + \lceil y \rceil - 1 \leq \lceil x + y \rceil \leq \lceil x \rceil + \lceil y \rceil$$

for any real numbers x, y.

Binomial Theorem For any positive integer n,

$$(a + b)^n = \sum_{k=0}^{n} \binom{n}{k} a^{n-k} b^k$$

where $\binom{n}{k} = \frac{n!}{k!(n-k)!}$ and $n! = 1 \cdot 2 \cdots n$ for any positive integer n. For any positive integer n, $\sum_{k=0}^{\lfloor n/2 \rfloor} \binom{n}{2k} = \sum_{k=0}^{\lfloor n/2 \rfloor} \binom{n}{2k+1} = 2^{n-1}$.

Sum and Product of Arithmetic Functions If $F(n) = \sum_{d \mid n} f(d)$, then $\sum_{i=1}^{n} F(i) = \sum_{i=1}^{n} \lfloor \frac{n}{i} \rfloor f(i)$. Similarly, if $F(n) = \prod_{d \mid n} f(d)$, then $\prod_{i=1}^{n} F(i) = \prod_{i=1}^{n} f(i)^{\lfloor \frac{n}{i} \rfloor}$.

Möbius Function and Möbius Inversion The Möbius function is

© The Editor(s) (if applicable) and The Author(s), under exclusive license to Springer Nature Singapore Pte Ltd. 2021
M. Billal and S. Riasat, *Integer Sequences*,
https://doi.org/10.1007/978-981-16-0570-3

$$\mu(n) = \begin{cases} 1 & \text{if } n = 1 \\ (-1)^{\omega(n)} & \text{if } n \text{ is square-free} \\ 0 & \text{otherwise} \end{cases}$$

where $\omega(n)$ is the number of distinct prime factors of n. For an arithmetic function $f : \mathbb{N} \to \mathbb{N}$, let $F(n) = \sum_{d|n} f(d)$. Then the Möbius inversion formula is

$$f(n) = \sum_{d|n} \mu(d) F\left(\frac{n}{d}\right)$$

Similarly, if $G(n) = \prod_{d|n} f(d)$, then

$$f(n) = \prod_{d|n} G\left(\frac{n}{d}\right)^{\mu(d)}$$

See Zeitz [Zei17, p. 239] and Billal and Hossein [BH18, Chap. 2, Sect. 5, Dirichlet Product and Möbius Inversion] for proof and further details.

Chinese Remainder Theorem   Let $n_1, n_2, \ldots, n_k$ be pairwise relatively prime integers and $a_1, a_2, \ldots, a_k$ be integers. Then there is a unique x modulo $n_1 n_2 \cdots n_k$ such that

$$x \equiv a_i \pmod{n_i}$$

If $N = n_1 n_2 \cdots n_k$ and $N_i = \frac{N}{n_i}$, then the solution x is given by

$$x \equiv \sum_{i=1}^{k} a_i M_i N_i \pmod{N}$$

where $M_i N_i \equiv 1 \pmod{n_i}$. See Billal and Hossein [BH18, Chinese Remainder Theorem, Sect. 5, Chap. 2] for a proof.

Chinese Remainder Theorem for Polynomials   Let $P_1(x), P_2(x), \ldots, P_k(x)$ be pairwise relatively prime integer polynomials and d_i be the degree of $P_i(x)$. If $A_1(x), A_2(x), \ldots, A_k(x)$ are polynomials such that $\deg(A_i) < d_i$, there is a unique polynomial $P(x)$ such that $\deg(P) < \sum_{i=1}^{k} d_i$ and

$$P(x) \equiv A_i(x) \pmod{P_i(x)}$$

If $\mathcal{P}(x) = P_1(x) P_2(x) \cdots P_k(x)$ and $\mathcal{P}_i(x) = \frac{\mathcal{P}(x)}{P_i(x)}$, then the solution is given by

$$P(x) \equiv \sum_{i=1}^{k} A_i(x) Q_i(x) \mathcal{P}_i(x) \pmod{\mathcal{P}(x)}$$

where $\mathcal{P}_i(x)Q_i(x) \equiv 1 \pmod{P_i(x)}$.

Bézout's Identity   For two integers a and b, there exist integers x, y such that $ax + by = \gcd(a, b)$. Moreover, if (x_0, y_0) is a solution to $ax + by = \gcd(a, b)$, then $(x_0 \pm bt, y_0 \mp at)$ is a solution for any integer t. Thus, we can without loss of generality assume that $x > 0, y < 0$ or $x < 0, y > 0$ if necessary by increasing or decreasing t. See Billal and Hossein [BH18, Bézout's lemma, Sect. 4, Chap. 2] for a proof.

Bézout's Identity for Polynomials   For two univariate integer polynomials $a(x)$ and $b(x)$ which do not have any common roots, there are polynomials $u(x)$ and $v(x)$ such that $a(x)u(x) + b(x)v(x) = 1$. For arbitrary integer polynomials $a(x)$ and $b(x)$, there are polynomials $u(x)$ and $v(x)$ such that $a(x)u(x) + b(x)v(x) = g(x)$ where $g(x)$ is the polynomial of largest degree that has only the common roots of both $a(x)$ and $b(x)$.

Vieta's Formulas   Let

$$f(x) = a_k x^k + a_{k-1} x^{k-1} + \cdots + a_1 x + a_0$$

be a polynomial and let $\alpha_1, \alpha_2, \ldots, \alpha_k$ be the roots of $f(x)$. Then we have the following identities:

$$\sum_i \alpha_i = -\frac{a_{k-1}}{a_k}$$

$$\sum_{i<j} \alpha_i \alpha_j = \frac{a_{k-2}}{a_k}$$

$$\vdots$$

$$\alpha_1 \alpha_2 \cdots \alpha_k = (-1)^k \frac{a_0}{a_k}$$

In short, $\sum \alpha_1 \alpha_2 \cdots \alpha_i = (-1)^i \frac{a_{k-i}}{a_k}$, where the sum is taken over all possible $\binom{k}{i}$ combinations of i roots taken at a time. This sum can also be written as an elementary symmetric polynomial in the roots $\alpha_1, \ldots, \alpha_i$. See Zeitz [Zei17, p. 199] for details.

Quadratic Residues   If $x \equiv a^2 \pmod{m}$ for some integer a, then x is a quadratic residue of m. For a prime p and a positive integer a, the Legendre symbol is defined as

$$\left(\frac{a}{p}\right) = \begin{cases} 0 & \text{if } p \mid a \\ 1 & \text{if } a \text{ is a quadratic residue of } p \\ -1 & \text{otherwise} \end{cases}$$

By Euler's criterion, this can also be defined using

$$\left(\frac{a}{p}\right) \equiv a^{\frac{p-1}{2}} \quad (\text{mod } p)$$

For two positive integers a and b, we have $\left(\frac{ab}{p}\right) = \left(\frac{a}{p}\right)\left(\frac{b}{p}\right)$. Gauss's law of quadratic reciprocity states that for two primes p and q,

$$\left(\frac{p}{p}\right)\left(\frac{q}{p}\right) = (-1)^{\frac{p-1}{2}\frac{q-1}{2}}$$

The Jacobi symbol is a generalization of the Legendre symbol. Let a be an integer and n have the prime factorization $n = p_1^{e_1} p_2^{e_2} \cdots p_r^{e_r}$, where the primes p_i are odd. The Jacobi symbol is defined as

$$\left(\frac{a}{n}\right) = \left(\frac{a}{p_1}\right)^{e_1} \left(\frac{a}{p_2}\right)^{e_2} \cdots \left(\frac{a}{p_r}\right)^{e_r}$$

And if $n = 1$, then $\left(\frac{a}{n}\right) = 1$. Similar to Legendre symbol, we have $\left(\frac{ab}{n}\right) = \left(\frac{a}{n}\right)\left(\frac{b}{n}\right)$ and if both m, n are odd then $\left(\frac{a}{mn}\right) = \left(\frac{a}{m}\right)\left(\frac{a}{n}\right)$. The law of quadratic reciprocity holds for the Jacobi symbol as well: for two odd integers m and n,

$$\left(\frac{m}{n}\right)\left(\frac{n}{m}\right) = (-1)^{\frac{m-1}{2}\frac{n-1}{2}}$$

See Billal and Hossein [BH18, Chap. 2, Sect. 8, Quadratic Residues] for proof and further details.

References

[BH18] M. Billal, A. Hossein, *Topics in Number Theory: An Olympiad-Oriented Approach*, 1st edn. (Amazon, 2018)
[Zei17] P. Zeitz, *The Art and Craft of Problem Solving* (Wiley, 2017)

Index

© The Editor(s) (if applicable) and The Author(s), under exclusive license to Springer   167
Nature Singapore Pte Ltd. 2021
M. Billal and S. Riasat, *Integer Sequences*,
https://doi.org/10.1007/978-981-16-0570-3

Lightning Source UK Ltd.
Milton Keynes UK
UKHW031134260622
404968UK00003B/25